The Joy of Mathematics

Sumita Bose

UNICORN BOOKS

Publisher

UNICORN BOOKS

F-2/16, Ansari Road, Daryaganj, New Delhi-110002

☎ 011-23275434, 23262683, 23250704 • *Fax:* 011-23257790

E-mail: info@unicornbooks.in • *Website:* www.unicornbooks.in

Branch : Mumbai

23-25, Zaoba Wadi, Thakurdwar, Mumbai-401002

☎ 022-22010941, 022-22053387

E-mail: rapidex@bom5.vsnl.net.in

ISBN 978-81-7806-232-7

The Joy of Mathematics

Edition: June 2015

Printed at : Param Offsetters, Okhla, New Delhi-110020

Preface

Mathematics has been an inseparable part of education since the beginning of formal education.

It has played an important role in the advancement of any civilization. It is a common belief that mathematics is dry, boring and a difficult subject. On the contrary if it is presented in an enjoyable manner, it is one of the easiest and interesting subjects.

In this book I have presented games, activities, magic and puzzles all based on mathematical concepts. I have also given amazing facts and patterns based on mathematics. I have had the opportunity to teach thousands of students of various levels over the past several years. Their love, respect, laughter, desire to learn, smiling faces and fear of mathematics have inspired me to write this book. I have provided several shortcut methods in this book for easy calculation.

Parents and teachers can use this book to supplement their teaching. I believe in the equation

Mathematics + Joy = Effective learning

When learning is fun both children and adults become interested in the learning process and are motivated to learn more. This book will provide that interest and motivation.

I avail this opportunity to convey my sincere thanks to Dr.Ashok Gupta and his team for their cooperation and untiring efforts in bringing out this book in an excellent form. I also thank my husband Mr.Parthasarathi Mitra for guiding me with the technical knowhow.

As there is always a scope for improvement, any suggestion from the readers are most welcome.

Sumita Bose

Salient Features

- Use of mathematics in everyday life
- Thought provoking puzzles
- Interesting mathematical magic, games and activities
- Amazing mathematical facts and patterns
- Tricks and shortcuts for quicker mathematics
- Hilarious mathematical jokes
- Simple and easy language
- Attractive picture illustrations
- Solved examples with every concept
- Practice exercise with answers

Contents

1

Mathemagic

a) Telling the Age:

Materials required:

Paper and pencil

Method:

1. Ask your friend to write down the number of the month in which he/she was born (Example: He/she has to write 9 if he/she was born in September).
2. Multiply the number by 2.
3. Add 5 to the product.
4. Multiply the total by 50.
5. Add the magic number of the year in which you are showing the magic (This number keeps changing, look at table 1.1).

Year	Magic Number
2009	1459
2010	1460
2011	1461
2012	1462

Table 1.1.

6. From step 5 subtract the year he/she was born in.
7. Ask him/her the final answer.
8. You can at once tell his/her age.

Example:

Suppose your friend was born in September 1999 and you are showing the magic in 2010 then he/she would work as follows:

1) 9 × 2 = 18
2) 18 + 5 = 23
3) 23 × 50 = 1150
4) 1150 + 1460 = 2610 (Magic number of year 2010 is 1460)
5) 2610 – 1999 = 611

Trick:

The last two digits of his/her final answer will be his/her age. In the above example your friend's age is 11(Last two digits of the final answer).

b) Thought Reader

Materials required:

None

CASE I

Method:

1. Ask your friend to think of any natural number.
2. Ask him/her to double the number.
3. Next ask him/her to multiply the number by 5 and tell you the answer.
4. You can look at his/her forehead and tell the number.

Example:

Suppose your friend thought 23 then he/she would work as follows:

1. Double of 23 = 46.
2. 46 × 5 = 230
3. He/She would tell you the final answer as 230.

Trick:

Remove the zero from the end and the remaining digits will form the number which your friend had thought. In the above example your friend thought 23(Remove the zero from the end).

CASE II

Method:

1. Ask your friend to think of any natural number.
2. Ask him/her to double the number.
3. Next ask him/her to add 30.
4. Finally divide by 2 and subtract the original number.
5. You can look at your friend's face and tell the answer.

Example:

Suppose your friend thought 17 then he/she would work as follows:

1. Double of 17 = 34.
2. 34 + 30 = 64.
3. 64 ÷ 2 = 32, 32 – 17 = 15

Trick:

Whatever you ask to add your friend's answer is half of that. In the above example half of 30 is 15.

c) Left or Right

Materials required:

Two pieces of paper and pencil

Method:

1. Take two small pieces of paper. Write 1 on the first paper and 10 on the second.
2. Give the two pieces to your friend. Ask him/her to keep one piece in his/her left hand and the other in his/her right and close his/her fist.
3. Next ask him/her to multiply the number in his right hand by 4 and the number in his/her left hand by 3.
4. Finally ask him/her to add the two products and tell you the answer.
5. You can tell which number was kept in the left hand and which one in the right hand.

Example:

Suppose your friend kept 10 in her right hand and 1 in her left hand then she would work as follows:

1. $10 \times 4 = 40$
2. $1 \times 3 = 3$
3. $40 + 3 = 43$.

Trick:

If the answer is odd, 1 is in the left hand and 10 in the right. If the answer is even it is vice versa.

d) Magic Number

Materials required:

Calculator, paper, envelope, pencil

Method:

1. Ask the person to enter the year of his/her birth in a calculator without showing it to you.
2. Ask him/her to first subtract 3 and then add his/her age.
3. On the date of the magic show if the person's birthday is yet to come ask him/her to add 1, if it's over he/she can skip this step.
4. Next ask him/her to subtract this number from the year of the magic show and then add 6.
5. Now ask him/her to square the number in his/her calculator.
6. Tell him to repeat step 5 two more time.
7. Now ask him/her to open the sealed envelope kept on the table. He/She will find his/her answer inside the envelope (Write the answer before starting the magic).

Example:

Suppose your friend's date of birth is 14th February 1999 and you are showing the magic on 14th March 2010 then your friend would calculate as follows:

1. 1999 - 3 = 1996, 1996 + 11 = 2007
2. He/She will skip step 2 since his/her birthday is over.
3. 2010 - 2007 = 3, 3 + 6 = 9
4. 9 × 9 = 81
5. 81 × 81 = 6561
6. 6561 × 6561 = 43046721

Trick:

The number is always 43046721.

e) Addition Wizard

Materials required:

Paper and two pencils

Method:

1. Ask your friend to write any five digit number with ones digit greater than 2.
2. You write a five digit number below your friend's number.
3. Repeat step 1 and step 2.
4. Then your friend writes another five digit number with ones digit greater than 2 and you write the total within seconds (without a calculator).

Example:

Friend's number	34567
Your number	65432
Friend's number	21062
Your number	78937
Friend's number	38549

Total	238547

Trick:

1. When you write the five digit number simply subtract each digit of your friend's number (34567 in this case) from 9. Your number 65432 in the second step is written as follows:

   9 – 3 = 6, 9 – 4 = 5, 9 – 5 = 4, 9 – 6 = 3, 9 -7 = 2.
2. In the same manner write your number (78937) in the fourth step (9 - 2 = 7, 9 - 1 = 8, 9 – 0 = 9, 9 – 6 = 3, 9 -2 =7).

3. To write the total start with a 2 (always), next write the first four digits of your friend's number in the fifth step. For the ones digit just subtract 2 from the one's digit of your friend's last number [in this case (9 - 2 = 7)].

Friend's number 34567

Your number 65432 (9–3 = 6, 9–4 = 5, 9–5 = 4, 9–6 = 3, 9-7 = 2)

Friend's number 21062

Your number 78937 (9 - 2=7, 9 – 1=8, 9 – 0=9, 9 – 6=3, 9 -2=7)

Friend's number 38549

Total 238547 → (Friends last number's unit's digit - 2)

Always 2 ← 2; 3854 ↓ First four digits of your friend's last number

f) Calendar Magic

Materials required:

Calendar, paper and pencil

January 2010

SUN	MON	TUE	WED	THU	FRI	SAT
31					1	2
3	4	5	6	7	8	9
10	11	12	13	14	15	16
17	18	19	20	21	22	23
24	25	26	27	28	29	30

Table 1.2(a)

Method:

1. Give a calendar to your friend and ask him/her to take out any month.
2. Ask him/her to select any 3 by 3 square in the calendar without showing it to you.
3. Next tell him/her to add all the numbers in the square and divide the sum by the number in the middle.
4. You can tell him/her the answer.

Example:

Suppose your friend selects the following 3 by 3 square.

5	6	7
12	13	14
19	20	21

Table 1.2(b)

Your friend will work as follows:

1. $5 + 6 + 7 + 12 + 13 + 14 + 19 + 20 + 21 = 117$
2. $117 \div 13$(middle number) $= 9$.

Trick:

The answer is always nine.

g) Number Sequence

Materials required:

Paper and pencil

Method:

1. Ask your friend to write any five consecutive numbers between 1 and 100.
2. Next ask him/her to add those numbers and tell you the total.

3. Just by knowing the total you can tell him/her the five numbers.

Example:

Suppose your friend chooses 25, 26,27,28,29 then his/ her total would be 135.

Trick:

Divide the total by 5. 135 ÷ 5 = 27. The quotient will give you the middle of the five numbers. You can easily find out the remaining four numbers.

h) Duplicate Numbers

Materials required:

Calculator

Method:

1. Ask your friend to think of any three digit number.
2. Ask him/her to repeat it and enter in a calculator.
3. Tell him/her to divide this six digit number by 7.
4. Now tell him/her to divide the answer by 11.
5. Finally tell him/her to divide by the three digit number he thought of.
6. You can tell your friend the answer.

Example:

Suppose your friend thought 259 then he/she would work as follows:

259259

259259 ÷ 7 = 37037

37037 ÷ 11 = 3367

3367 ÷ 259 = 13

Trick:

The answer is always 13. If you want 11 as the answer then tell your friend to divide by 13 instead of 11 in the fourth step.

i) Fibonacci Trick

Leonardo Fibonacci was an Italian mathematician who replaced the Roman numerals by Arabic numbering system in Europe. He also developed a sequence of numbers which now bears his name.

Fibonacci sequence:

It starts with 1. Each subsequent number is the sum of two previous numbers. The first few numbers of this sequence are:

1	**1**	**2**	**3**	**5**	**8**	**13**	**21**
		1+1=2	1+2=3	2+3=5	3+5=8	5+8=13	8+13=21

Fig. 1.1

Materials required:

Paper, pencil and ruler

Method:

1. Make ten rows on a sheet of paper and ask one friend to write any two natural numbers on the first two rows (One number in each row).
2. Now ask him/her to add the two numbers and write the sum in the third row.
3. Ask him/her to continue the series, adding the last two numbers together each time.
4. When your friend reaches row 10, ask him/her to add all the numbers together.
5. Turn over your answer to show that you already had the answer.

Example:

Suppose your friend chooses two numbers as 3478 and 5690 then he/she would work as follows:

Row	Number
Row 1	3478
Row 2	5690
Row 3 (Row 1 + Row 2)	9168
Row 4 (Row 2 + Row 3)	14858
Row 5 (Row 3+ Row 4)	24026
Row 6 (Row 4 + Row 5)	38884
Row 7 (Row 5 + Row 6)	62910
Row 8 (Row 6 + Row 7)	101794
Row 9 (Row 7 + Row 8)	164704
Row 10 (Row 8 + Row 9)	266498

Table 1.3

3478 + 5690 + 9168 + 14858 + 24026 + 38884 + 62910 + 101794 + 164704 + 266498 = 692010

Trick:

When your friend reaches the seventh row, quickly look at the paper, and then multiply this number by 11 (62910 × 11 = 692010). Write the number on a piece of paper and turn it upside down on the table. (Property of Fibonacci sequence: Eleven times the seventh number of the Fibonacci sequence is equal to the sum of first ten numbers of the sequence).

j) Addition with Closed Eyes

Materials required:

Six rectangular pieces of paper, sketch pen, scarf or a big handkerchief

Method:

1. In this magic you need six papers. Take the first paper; write 1 in front and 2 at the back. Similarly take the second paper and write 3 in front and 4 at the back. Write 5 and 6 on the third paper, 7 and 8 on the fourth, 9 and 10 on the fifth and 11 and 12 on the sixth paper(To make it more interesting write the numbers with a sketch pen on a thin cardboard covered with paper).
2. Hand over the papers to your friend.
3. Ask him/her to tie your eyes with a scarf or handkerchief.
4. Now ask him/her to shuffle the papers and display them on the table. He/she can have only odd, only even or a mixture of even and odd numbers. Tell him/her to find the total of the numbers displayed.
5. Finally ask him/her to tell you the number of odd numbers displayed and you can at once tell the total.

Example:

Suppose your friend displays 2, 3, 5, 7, 10, 12(three odd numbers displayed) then his/her total would be 2 + 3 + 5 + 7 + 10+ 12 = 39.

Trick:

Subtract the number of odd numbers from the magic number 42(In this case 42 – 3 = 39).

k) Magic Product

Materials required:

Calculator

Method:

1. Ask your friend to multiply 142857 by any number from 1 to 7.

2. Next ask him/her to tell you any one digit and the position of the digit in the answer.
3. You can tell the complete answer just by knowing one digit.

Example:

Suppose your friend multiplies 142857 by 6 then his/her answer would be 857142.

Trick:

Every product in this magic will have the same series of digits (142857). It will always be in a circular manner. In the above example if your friend says 8 in the first position from left you write 857 first and then put 142 in the end. Suppose your friend says 5 in the fifth position from left, you write 57 and put 1428 in front. Thus the answer is 142857.

In case your friend says 9 in any position the answer will be 999999(142857 × 7 = 999999).

I) Two Digit Numbers

Materials required:

Paper and pencil

Method:

1. Ask your friend to write 2 two digit numbers.
2. Ask him/her to multiply the first by 5 and add 36.
3. Next ask him/her to multiply the result by 20.
4. Finally ask him/her to add the other two digit number and subtract 720.
5. You can tell your friend's numbers just by looking at the final answer.

Example:

Suppose your friend chooses 19 and 23 then he/she would work as follows:

1. $19 \times 5 = 95$
2. $95 + 36 = 131$
3. $131 \times 20 = 2620$
4. $2620 + 23 = 2643$
5. $2643 - 720 = 1923$.

Trick:

The first two digits of the final result will represent the first number thought by your friend and the last two digits will represent the second number thought by your friend.

m) Brothers, Sisters and Friends

Materials required:

Paper and pencil

Method:

1. Ask your friend to write the number of his/her sisters and multiply it by 2.
2. Next add 3 to that total.
3. Now multiply the result by 5 and add the number of brothers to this answer.
4. Further multiply this answer by 10 and add the number of very good friends.
5. Finally add 125 and tell you the result.
6. You can tell the number of sisters, brothers and friends.

Example:

Suppose your friend has 3 sisters, 2 brothers and 4 very good friends then he/she would work as follows:

1. $3 \times 2 = 6$
2. $6 + 3 = 9$
3. $9 \times 5 = 45$, $45 + 2 = 47$
4. $47 \times 10 = 470$, $470 + 4 = 474$
5. $474 + 125 = 599$

Trick:

You subtract 275 from the final answer. The result will give you the number of sisters, brothers and friends respectively. $599 - 275 = 324$.

Number of sisters → 324 ← Number of friends

↓

Number of brothers

n) Card Magic

Materials required:

A pack of cards

Suit number chart:

♣ = 1   ♦ = 2   ♥ = 3   ♠ = 4

Fig. 1.2

Card number chart:

Ace = 1, Jack = 11, Queen = 12, King = 13.

Method:

1. Keep a pack of cards on the table.
2. Ask your friend to pick up any card from the deck and carefully see the number as well as its suit.
3. Now ask him/her to multiply the card number by 2 and add 2 to the answer.
4. Next multiply the answer by 5.
5. Finally add the suit number and tell you the result.
6. You can look at him/her and tell the card picked by him/her.

Example:

Suppose your friend picks 7 of spades then he/she would work as follows:

1. 7 × 2 = 14, 14 + 2 = 16
2. 16 × 5 = 80
3. 80 + 4 = 84(Suit number of spade is 4)

Trick:

Subtract 10 from the final answer(84 - 10 = 74). The left hand digit will give you the card number and the right hand digit will give you the suit corresponding to the suit number chart.

2

Mathematical Games

a) Number Tic Tac Toe

Number of players:

Two

Materials required:

Two different coloured sketch pens or pencils, paper

Rules:

1. One player takes the one digit odd numbers (1, 3, 5, 7, and 9) and the other player takes the one digit even numbers (0, 2, 4, 6, and 8).
2. The player with the odd number starts the game.
3. The players have to put their numbers in the 3 by 3 grid one at a time.
4. Each number can be used only once.
5. The winner is the player who completes a line (horizontal, vertical or diagonal) of three numbers which add up to 15.

Example:

First player – Odd,   Second player - Even

		4
	3	

Fig. 2.1(a)

First move – First player puts 3 and Second player puts 4.

		4
5	3	6

Fig. 2.1(b)

Second move – First player puts 5 and Second player puts 6.

9		4
5	3	6

Fig. 2.1(c)

Third move – First player puts 9. Now Second player can either put 2 in between 9 and 4 (horizontal total 15) or 8 below 5(diagonal total 15). In either case blue wins.

b) Quilt Design

Number of players:

Two to four

Materials required:

Paper, as many different coloured sketch pens or pencils as the number of players, ruler

Rules:

1. Each player colours a small triangle in the following grid on his/her turn (photocopy the grid or draw it on a paper).

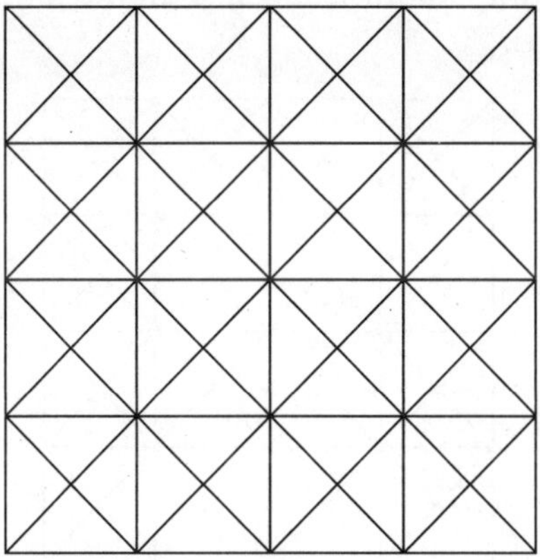

(a)

1. They alternate turns and can colour anywhere on the grid.
2. The points are awarded according to the following rule:

Square:	1 Point
Triangle:	2 Points
Rectangle:	3 Points
Parallelogram:	4 Points

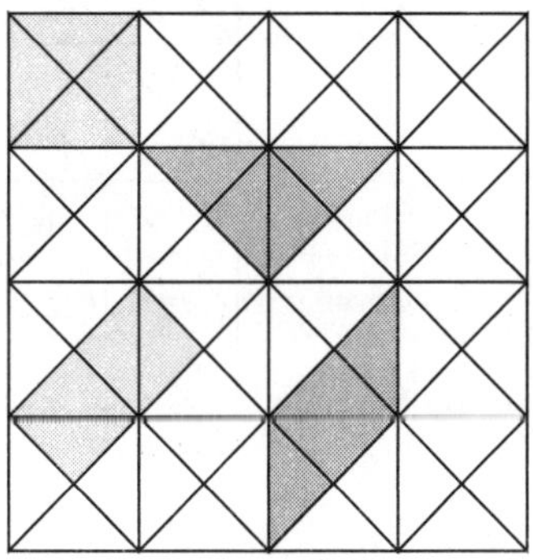

(b)

3. The game is over when the grid is completely coloured.
4. The player with the maximum points wins the game.

Example:

First Player – Dark, Second Player – Light

First move: The player with dark colour plays first. The player with light colour scores 1 point since he/she has made a square.

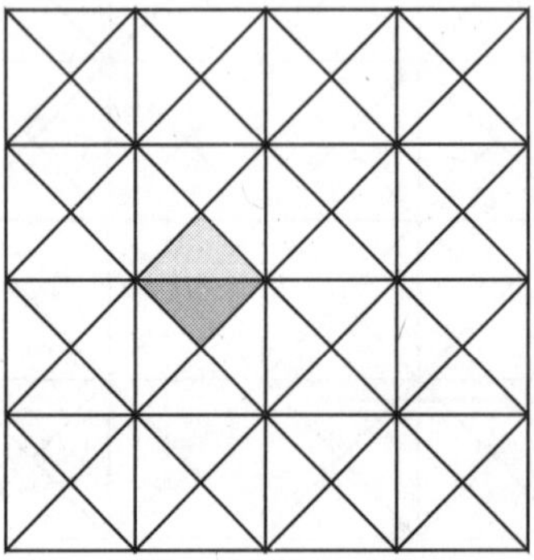

Fig. 2.1(c)

Second move: The first player scores 2 points since he/she has made a triangle in the third row. The second player scores 3 points for making a rectangle.

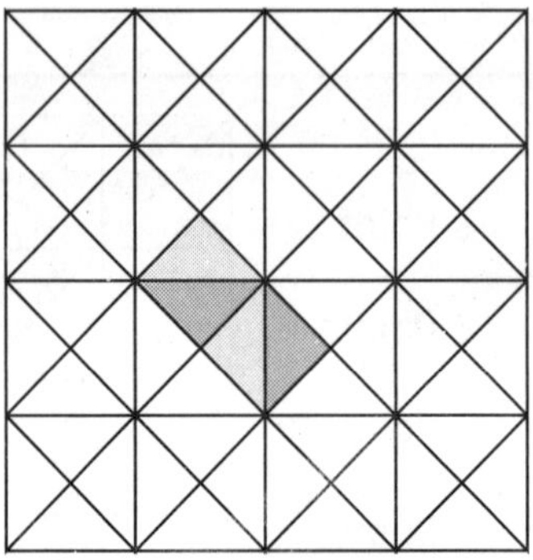

Fig. 2.1(d)

The game continues in this manner till the grid is completely coloured.

c) Pentafun

Number of players:

Any

Materials required:

Paper, pen, ruler and nine counters (or buttons or coins)

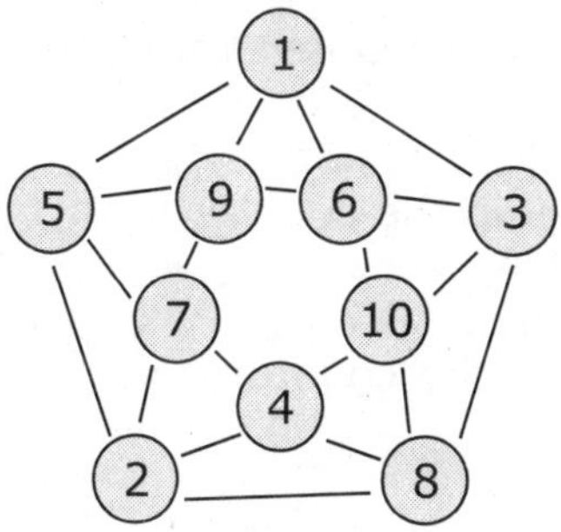

Fig. 2.2.

Rules:

1. This game is played one at a time by the players.
2. Start by placing nine counters or buttons or coins on any of the ten spots leaving one spot vacant.
3. Jump one counter (or button or coin) over another along the diagonal to reach the empty space. Remove the jumped counter. For example if 6 is vacant and you have counters on all other numbers you may jump the counter from 5 to 6 by removing 9 or jump the counter from 8 to 6 by removing 10. However you cannot slide from 3 to 6.
4. Slide along the sides to reach the vacant spot. For example if 3 is vacant you may slide from 8 to 3 or 1 to 3.
5. Continue sliding or jumping and removing counters until it is not possible to jump or slide any more.

6. Add up the numbers on the empty spots.
7. The person with the greatest total is the winner.

Example:

Suppose the player puts his/her counters on every number except 9.

He/she plays as follow:

First move: Jump the counter from 2 to 9 and remove the counter from 7.

Second move: Jump the counter from 1 to 7 and remove the counter from 9.

Third move: Jump the counter from 10 to 2 and remove the counter from 4.

Fourth move: Jump the counter from 5 to 4 and remove the counter from 7.

Fifth move: Slide the counter from 3 to 1.

Sixth move: Jump the counter from 8 to 7 and remove the counter from 4.

Seventh move: Jump the counter from 2 to 9 and remove the counter from 7.

Eighth move: Jump the counter from 9 to 3 and remove the counter from 6.

Ninth move: Slide the counter first to 8 and then to 2.

Now add up the numbers on empty spots. So the score is 52.

d) Nim

This game is said to have originated in China where it was known as "Jianshizi" or "Picking stones". A similar game was also mentioned by Luca Pacioli an Italian mathematician. Its current name("nimm" in German means " take") was coined by the mathematics professor Charles Bouton of Harvard University in 1901.

Number of players:

Any

Materials required:

Sixteen match sticks

Fig.2.3

Rules:

1. In this game the players take alternate turns.
2. The match sticks are arranged as shown above.
3. The players take turns removing any number of matches from a single row.
4. The player who picks up the last stick wins the game.

Example:

First Player – Removes three sticks from the second row.

Second Player – Removes three sticks from the fourth row.

First Player – Removes one stick from the first row.

Second Player – Removes two sticks from the fourth row.

First Player – Removes three sticks from the third row.

Second Player – Removes two sticks from the fourth row.

First Player – Removes two sticks from the third row and wins the game.

e) Alquerque

This game is also known as "Quirkat". It is said to have originated in the Middle East around 1400 B.C. It is the parent game of modern day "checkers".

Number of players:

Two

Materials required:

Paper, pencil, ruler, 12 counters (or buttons) of one colour, 12 counters (buttons) of another colour

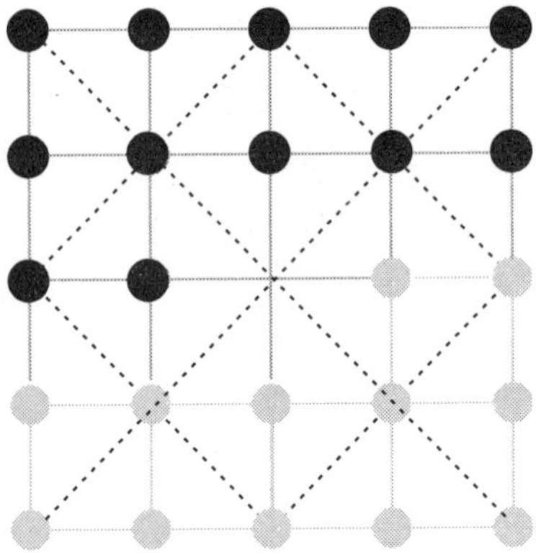

Fig. 2.4(a)

Rules:

1. In this game the players take alternate turns.
2. A board (5 by 5) is drawn (or photocopied) like above. The lines indicate the moves allowed.

3. Each player has twelve counters arranged as above.
4. The objective of this game is to take as many of the opponent's counters as you can or to create a situation in which the opponent is unable to move.
5. The players can make one of the following moves:

Ordinary move - Moving the counter along a line to an adjacent vacant position.

Capturing move – Hopping opponent's one counter along the line thereby taking the opponents counter.

6. When no further moves are possible the player with the maximum number of opponent's counters is the winner.

Example:

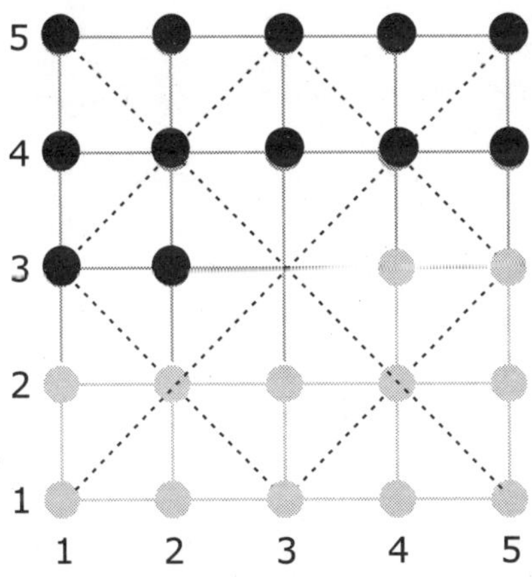

Fig. 2.4(b)

The first player makes an ordinary move by shifting his/her counter from (2,3) i.e. 2nd column, 3rd row to (3, 3). The second player makes a capturing move by hopping his/her counter from (4,3) to (2,3) thereby taking the first players counter. The game continues till there are no possible moves. The player with the maximum number of opponent's counters wins the game.

f) Pong Hau K'i

This is an ancient Chinese game. In Korea it is known as "Kono". It is somewhat similar to "tic tac toe."

Number of players:

Two

Materials required:

Paper, pencil, ruler, 2 counters (or buttons) of one colour, 2 counters(buttons) of another colour

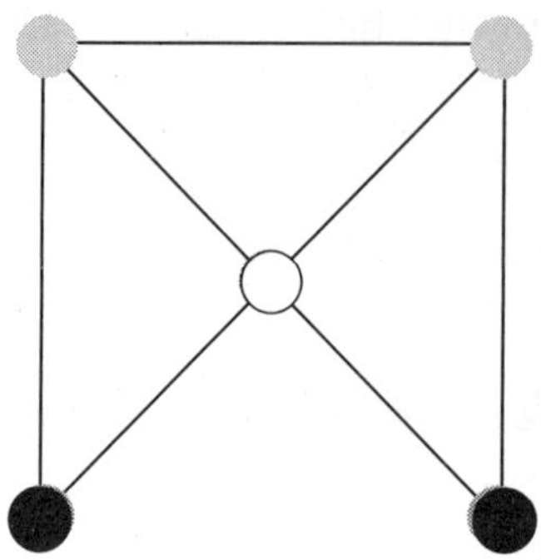

Fig. 2.5(a)

Rules:

1. In this game the players take alternate turns.
2. A board is drawn (or photocopied) like above. The lines indicate the moves allowed.
3. Each player has two counters arranged as above.
4. The objective of the game is to trap the opponent so that no moves are available for him/her.
5. Each player can move his/her counter to the adjacent space by moving along the line.
6. When no further moves are available the player who blocked the opponent wins the game. If the same set

of moves is repeated four times the game ends in a draw.

Example:

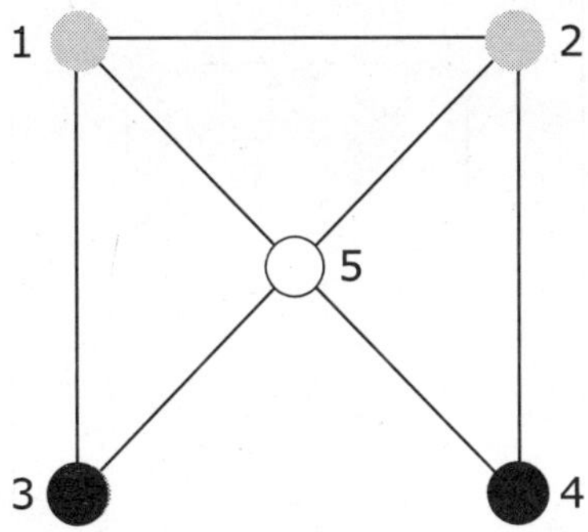

Fig. 2.5(b)

The first player moves his/her counter from 1 to 5. The second player moves from 3 to 1.

The game continues by moving along a line till no further move is available.

g) Nine Men Morris

This game is found to be carved into an Egyptian temple Kurna(around 1440 B.C). European boards are found in the first city of Troy. Shakespeare mentioned about this game in his "A Midsummer Night's Dream."

Number of players:

Two

Materials required:

Paper, pencil, ruler, 9 counters(or buttons) of one colour, 9 counters(buttons) of another colour

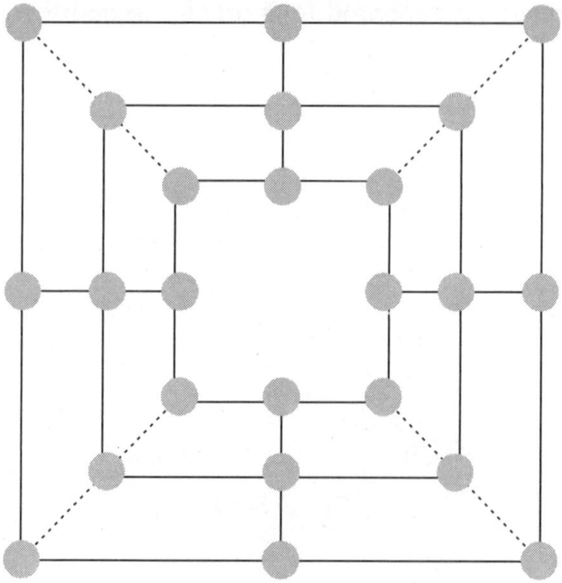

Fig. 2.6(a)

Rules:

1. In this game the players take alternate turns.
2. A board is drawn (or photocopied) like above. The lines indicate the moves allowed.
3. Each player has nine counters to start the game. They can arrange the counters on any circle(Placing three counters in a row is not allowed).
4. The objective of the game is to trap the opponent so that no moves are available for him/her or he/she is left with just two counters.
5. The players can move any of their counters to an adjacent vacant circle. Whenever a player forms three counters in a row(this is called mill) he/she can take any of the opponents counter that is not in any mill.(If all the counters are in a mill it is treated as an exception and any counter is taken.)
6. When no further moves are available the player who blocked the opponent wins the game.

Example:

The two players arrange their counters in the following manner:

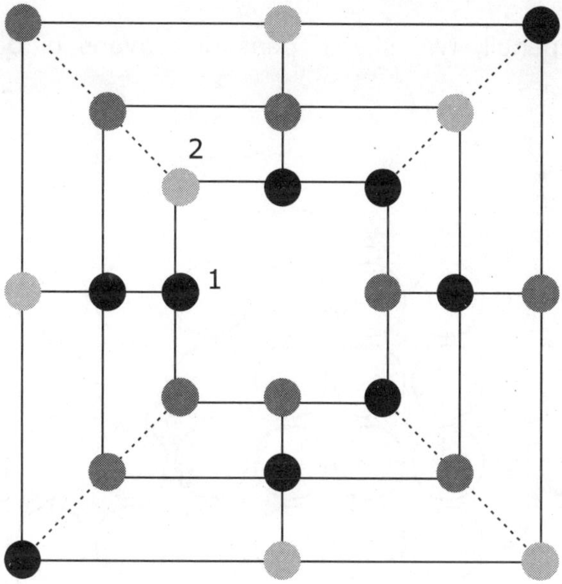

Fig. 2.6(b)

The player with the black counter moves his/her counter from position 1 to position 2 thereby making a mill (three counters in row). He/she can take any of the opponents counter. The game continues till no further move is possible. The player blocking the opponent is the winner.

h) Captain's Mistress

This game was played by Captain Cook on his long voyages. It entertained him and his crew. He spent so much time playing this game at night that it came to be known as "Captain's Mistress". However at present it is known as "Connect Four" under the trademark of Milton Bradley(1974).

Number of players:

Two

Materials required:

Paper, pencil, two sketch pens or crayons of different colour

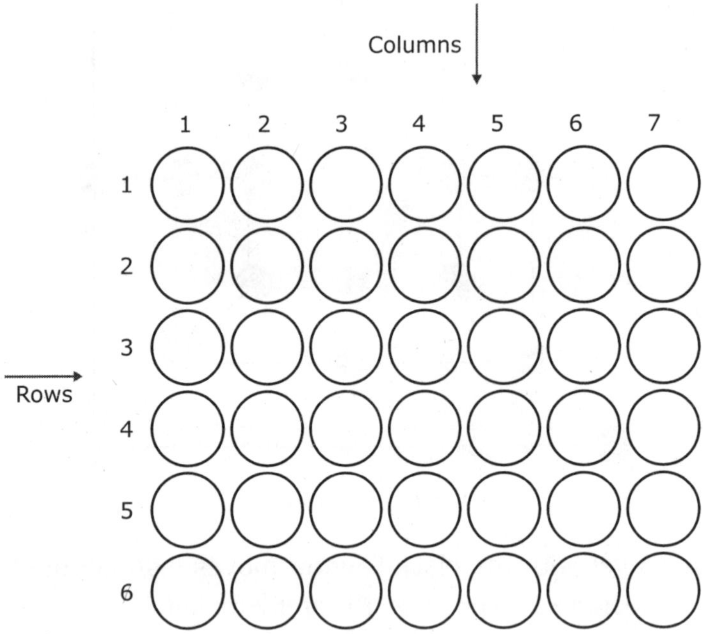

Fig. 2.7(a)

Rules:

1. In this game the players take alternate turns.
2. This game is normally played on a board with slits where the counters can be inserted. It may also be played by drawing(or photocopying) the above board.
3. The two players take different colour pencils or crayons.
4. Each player has to colour a circle which can be in row six or adjacent to any previously coloured circle.

5. The objective of the game is to colour four circles in a row, column or diagonal and to stop the opponent from achieving this.

6. The winner is the player who can colour four circles in a row, column or diagonal first. If all forty two circles are coloured and no one gets four circles then the game ends in a draw.

Example:

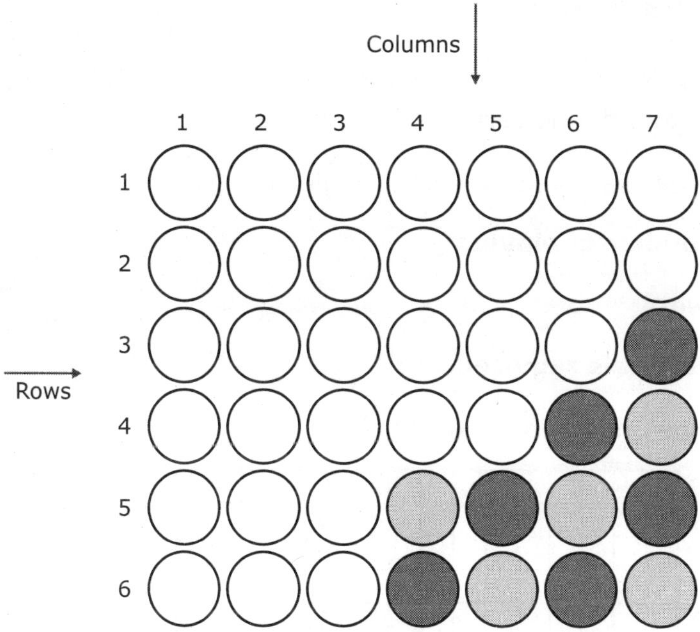

Fig. 2.7 (b)

First player : Dark colour, Second player: Light colour

First move: The first player colours (4,6) i.e 4th column, 6th row. The second player colours (5,6).

Second move: The first player colours (5,5). The second player colours (7,6).

Third move: The first player colours (6,6). The second player colours (6,5).

Fourth move: The first player colours (7,5). The second player colours (4,5).

Fifth move: The first player colours (6,4). The second player colours (7,4).

Sixth move: The first player colours (7,3) and wins the game.

i) Battleship

This is an old Navy game. It was invented by Clifford Von Wickler during early 1900.

It was patented by Milton Bradley in 1943.

Number of players:

Two

Materials required:

Paper, pencil, ruler, graph paper(optional)

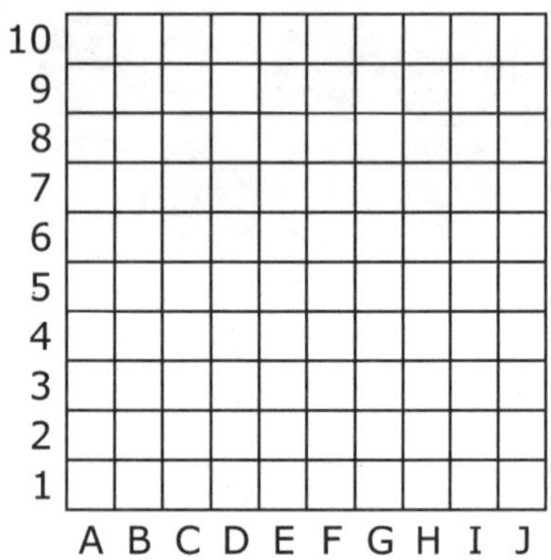

Fig. 2.8(a)

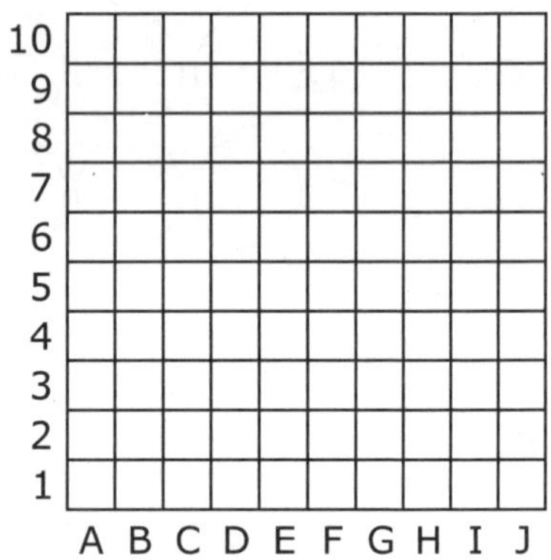

Fig. 2.8(b)

Rules:

1. In this game the players take alternate turns.

2. Each player draws (or photocopies) two grids like above(you may also use a graph paper). One for drawing own's ship and the other for keeping a record of the opponents "hits" and "misses".

3. Each player draws his/her ships into the grid according to the following rule:

   Aircraft carrier (5 blocks)

   Battleship (4 blocks)

   Cruiser (3 blocks)

   Submarine (3 blocks)

   Destroyer (2blocks)

   The ships can be drawn vertically or horizontally (They cannot overlap). It should be hidden from the opponent.

4. The objective of the game is to be the first player to sink each other's hidden ships.

5. Each player calls out one coordinate by a letter and a number. For example if one of the players say B3 and the opponent's one of the ships lies in that square the opponent will say "hit". If it doesn't, he/she will say "miss". The ship sinks when all the squares of that ship are hit.

6. A player wins if he/she can sink all his/her opponent's ships.

Example:

Suppose one of the player draws his/her ships according to the following:

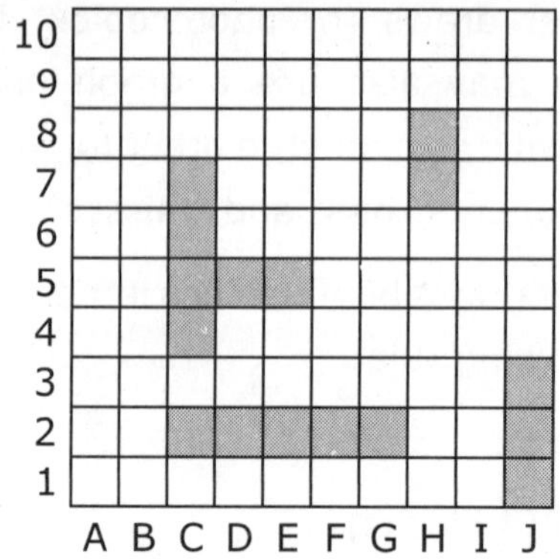

Fig.2.8 (c)

Now if the second player calls B2 the first player will say miss but if he/she calls C2 the first player will say hit. The game continues till all the squares of various ships are hit.

A particular ship sinks if all its squares are hit. For example, the aircraft carrier will sink if C2, D2, E2, F2 and G2 are hit. The player who can sink the opponents ships first is the winner.

j) Sprouts

The game of Sprouts was invented in 1967 by two mathematicians Michael S. Paterson and John H. Conway Cambridge University (UK).

Number of players:

Two

Materials required:

Paper and pencil

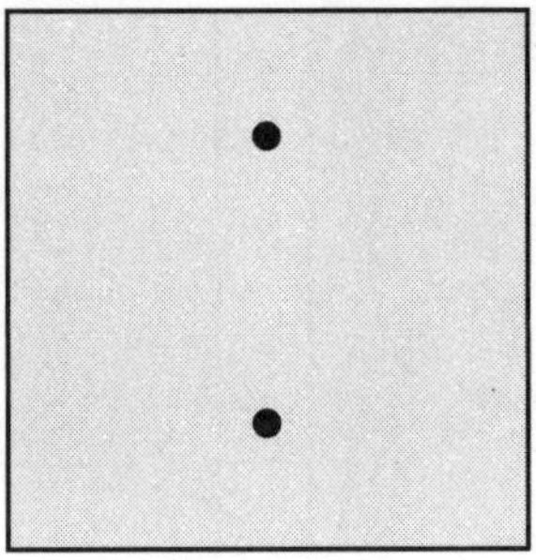

Fig.2.9(a)

Rules:

1. In this game the players take alternate turns.
2. To start the game any number of dots are drawn on a sheet of paper.
3. Each player has to draw a line(straight or curved) joining any two dots and putting a dot anywhere on the line. The line drawn cannot cross another line or itself. The line may start and end at the same point. However one point cannot have more than three lines attached to it.

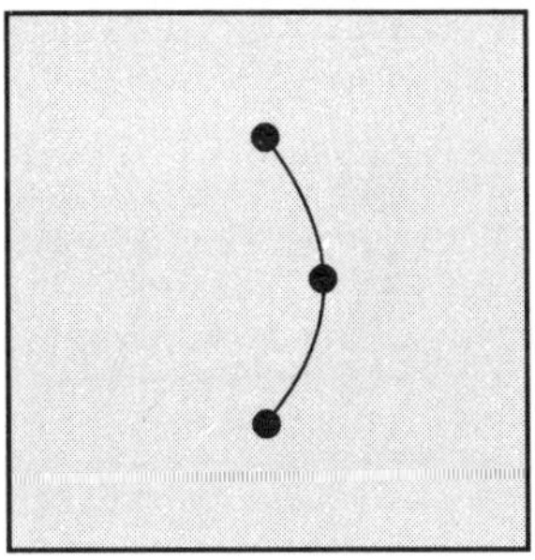

Fig. 2.9(b)

4. The objective of the game is to create a situation in which it is impossible for the opponent to draw any line.

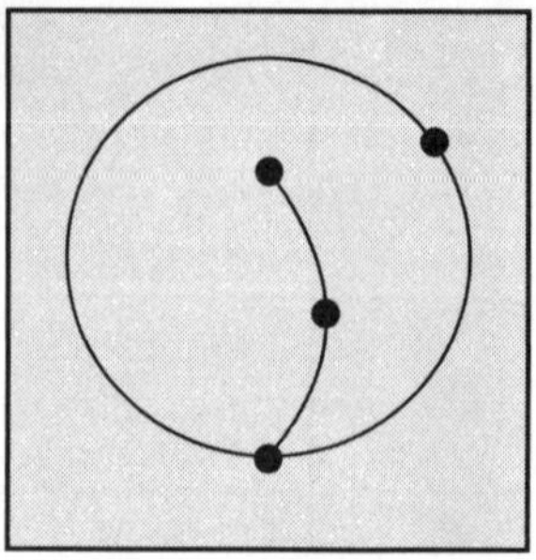

Fig.2.9(c)

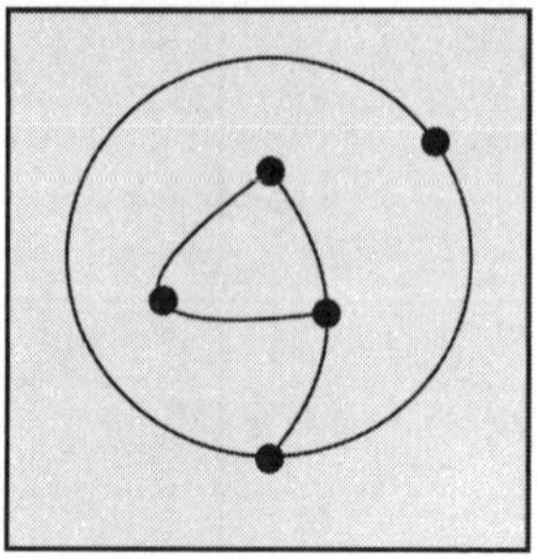

Fig.2.9(d)

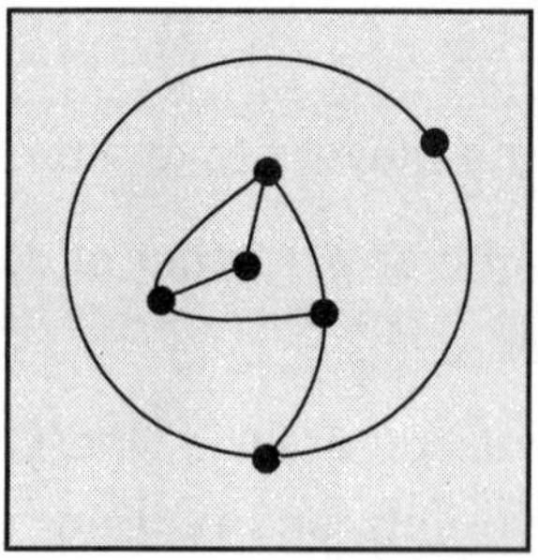

Fig.2.9 (e)

5. The last player to draw a line is the winner.

+++

3

Mathematical Activities

a) Permutation

Materials required:

Six equal sized paper, pencil, ruler, colours or coloured pictures (6), a pair of scissors, stapler and thick paper.

(a)

(b)

(c)

(d)

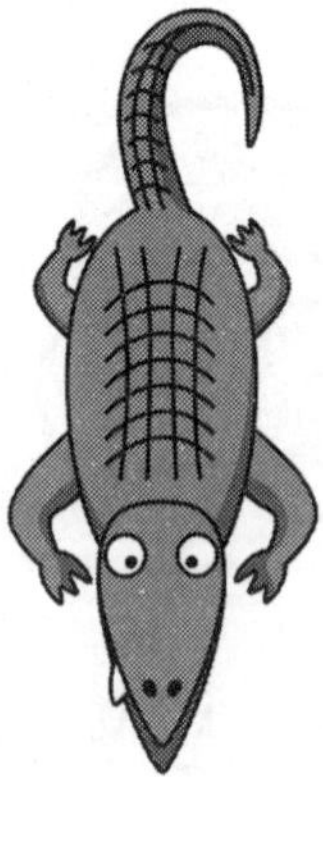

(e)

(f)

Fig. 3.1

Method:

1. Cut out six equal sized papers. Draw or paste pictures on them.

2. Draw two horizontal lines on each picture approximately at equal distance.

3. Put all the six pages on top of each other and staple them from one side.
4. You may use some thick paper (chart paper) as the first and the last page.
5. Now you have a thin booklet.
6. Cut out all the pictures along the horizontal lines. (Do not cut the cover pages)
7. Open the pages at random and see how many different permutations of animals you get.

b) Paper Folding

Materials required:

One long strip of paper, one ordinary paper, ruler, pencil

Method:

1. Take a long strip of paper and fold it once.
2. Keep folding till it is possible.
3. Record your observation in the following table.

Number of times folded	Number of Sections	Number of creases

Table 3.1

If it were possible to fold the paper nine times, how many sections and how many creases would you get?

c) Geometric Designs

Materials required:

Paper, pencil, compass, ruler, eraser

Method:

1. Draw a circle of radius 9 cm.
2. Mark any point P on the circle. Without changing the radius of the circle keep the compass point at P and mark an arc Q.
3. Next keep the compass point at Q and mark an arc R.
4. Continue the process till you reach S.

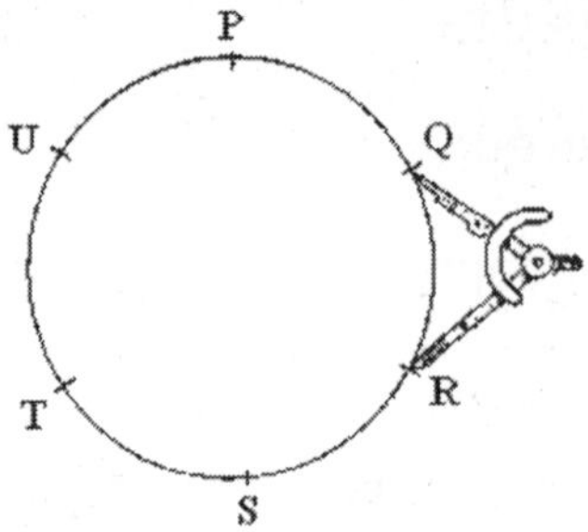

(a)

5. Now use your ruler to draw the following basic pattern. (Draw it three times.)

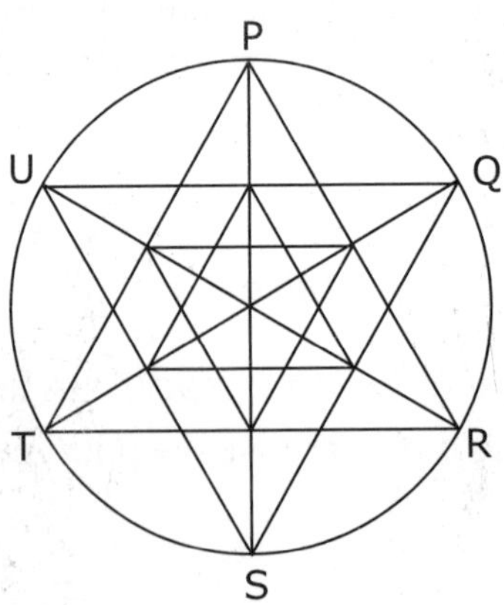

(b)

6. Rub out some lines from the above basic pattern to draw the following geometric designs.

(c)

(d)

(e)

FIg. 3.2

d) Regular Polyhedra

A regular polyhedron is a polyhedron with congruent faces and identical vertices. Although there are many polyhedra but there are only five regular polyhedra. They are collectively known as the Platonic solids (named after Plato, the famous Greek philosopher and mathematician).

Materials required:

Paper, pencil, ruler, tracing paper, thin cardboard, glue

Method:

1. Trace the following net diagrams.

2. Paste them on a thin cardboard.

3. Fold along the lines, form the shapes of the platonic solids and paste the edges with glue. Then fill in the following table:

Shapes	No. of Faces	No. of Edges	No. of Vertices

Table 3.2

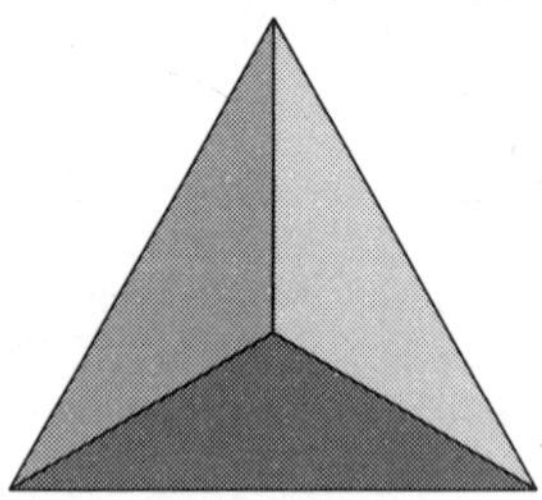

(a) Tetrahedron

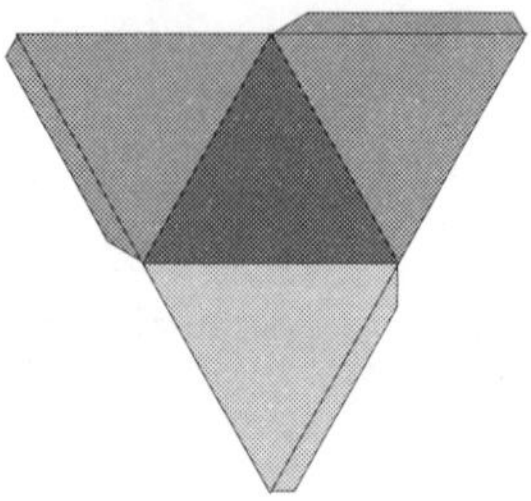

(b) Net diagram of tetrahedron

Fig.3.3

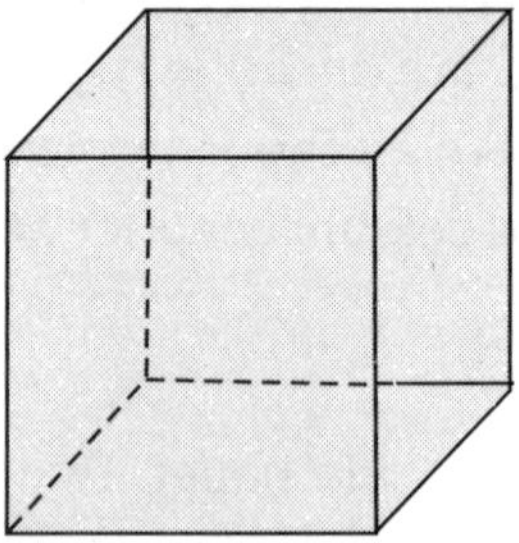

(a) Cube

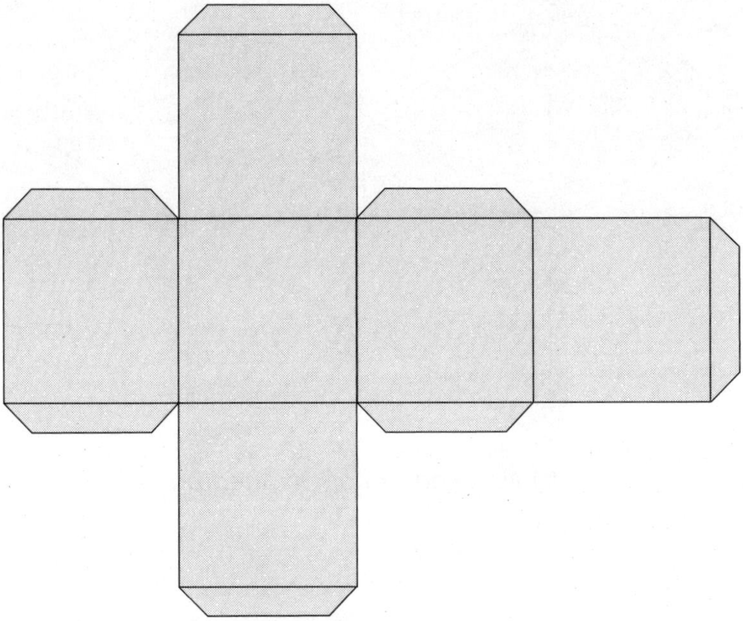

(b) Net diagram of cube

Fig.3.4

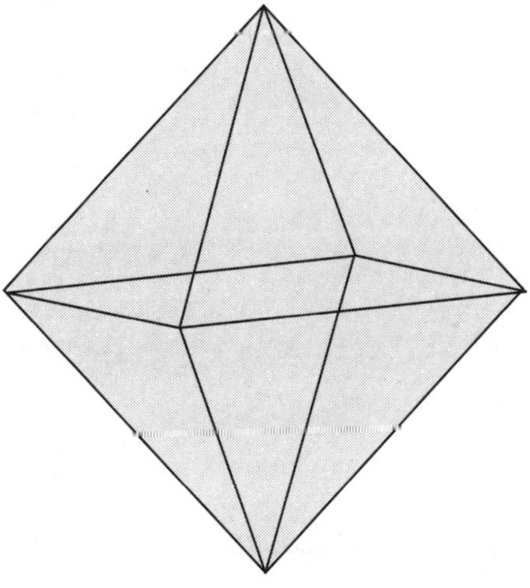

(a) Octahedron

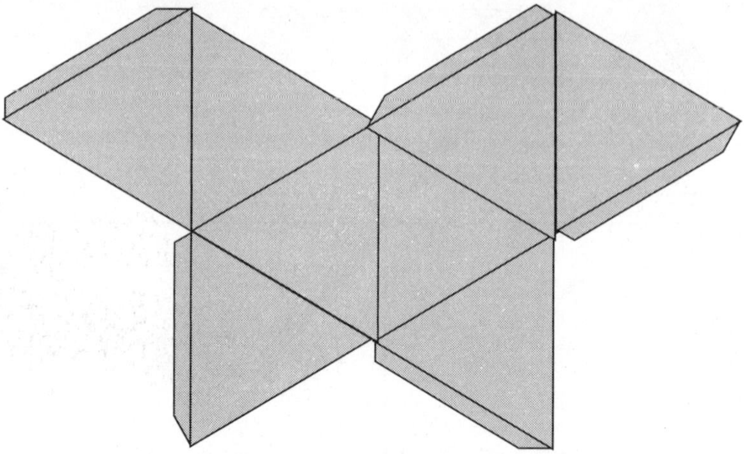

(b) Net diagram of Octahedron

Fig.3.5

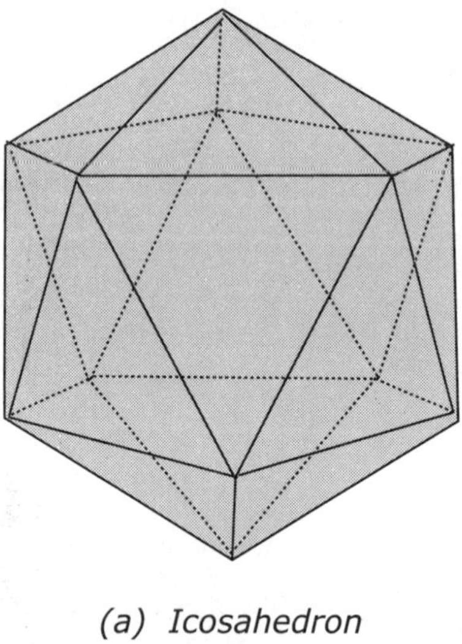

(a) Icosahedron

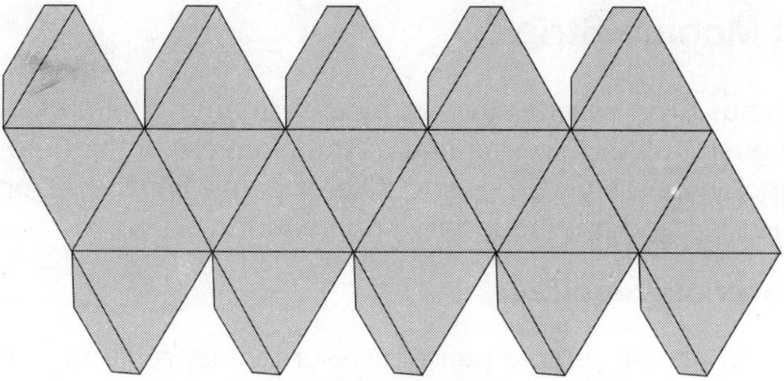

(b) Net diagram of Icosahedron

Fig.3.6

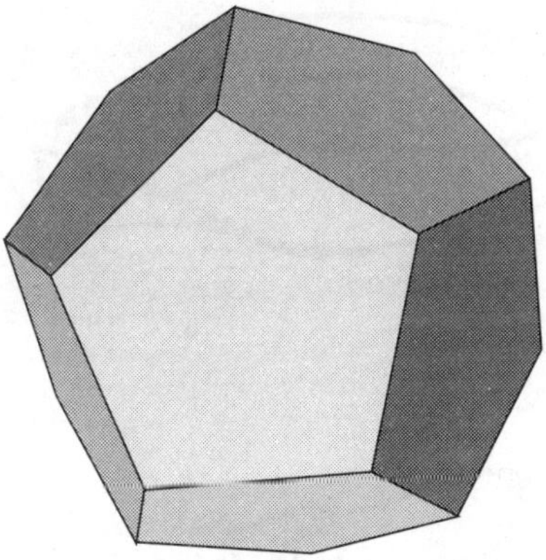

(a) Dodecahedron

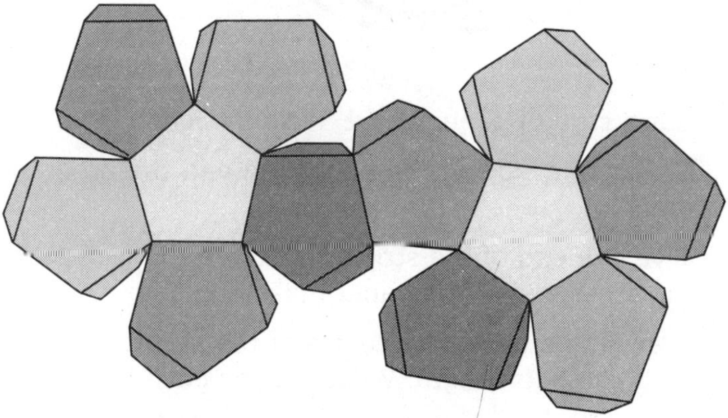

(b) Net diagram of dodecahedron

Fig.3.7

e) Möbius Strip

Möbius Strip was discovered by a German mathematician, August Möbius. He was one of the pioneers in the field of topology. Möbius Strip is a strip of paper which has only one side and one boundary component.

Materials required:

Paper, pencil, ruler, a pair of scissors, glue, red and green crayons

Method:

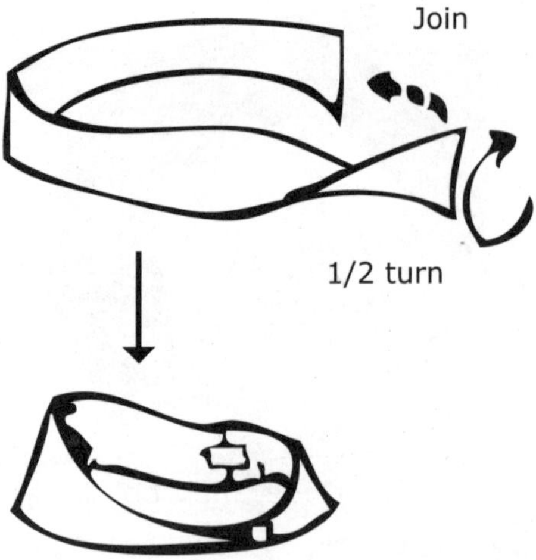

Fig.3.8

1. Cut a strip of paper about 30cm by 5cm.
2. Colour one side of the paper red and the other side green.
3. Give one end of the strip a half twist and glue to it to the other end of the paper making a loop.
4. Draw a line along the centre of the strip and cut the strip along its length. What did you get?
5. Moebius strips are often used as belts in machinery because they last longer.

f) Pythagoras Theorem

Materials required:

Paper, pencil, compass, ruler, eraser, graph paper

Method:

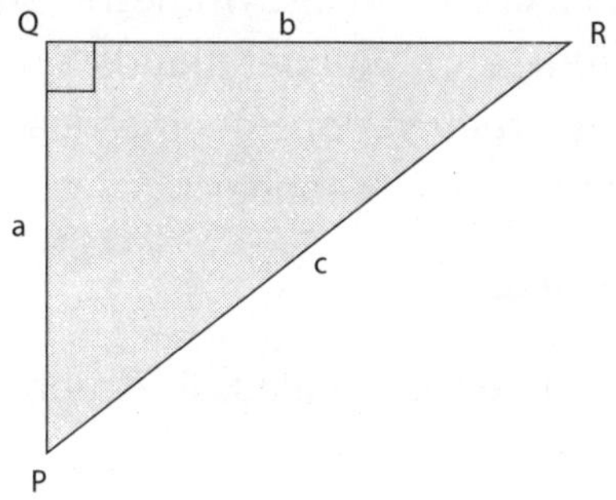

Fig. 3.9

1. Draw a right angled triangle with perpendicular 3cm, base 4cm and hypotenuse 5cm. (In the above diagram PQR is a right angled triangle right angled at Q. The hypotenuse PR measures *c*, PQ measures *a* and QR measures *b*.)
2. With side lengths 5cm, 4cm and 3cm draw three separate squares on a graph paper.
3. Count the number of small squares in each of the squares and record your observation.
   a) Total small squares in a square of side 5cm = ________
   b) Total small squares in a square of side 4cm = ________
   c) Total small squares in a square of side 3cm = ________
   d) Total small squares in a square of side 3cm + Total small squares in a square of side 4cm = ________

What can you conclude? Try with other kind of triangles. Do you get the same conclusion?

g) Geoboard

A geoboard is a wooden board in which certain number of nails are half driven in a symmetric square (five by five, four by four etc). Rubber bands are used around the nail heads to study various geometrical concepts. The following is a five by five geoboard.

Materials required:

Cardboard box, nine thumb pins, 6-7 rubber bands

Method:

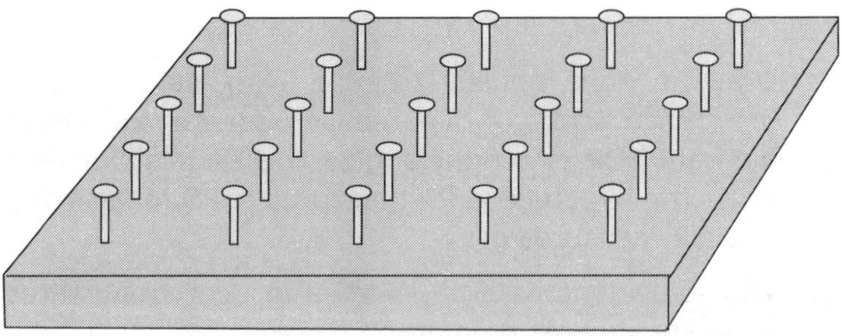

(a)

Case I

Make a three by three geoboard (it can also be made on a cardboard box using thumb pins). How many different squares can you make on this board?

Case II

Finding area on a geoboard

Let us make the following shape on a geoboard and find its area.

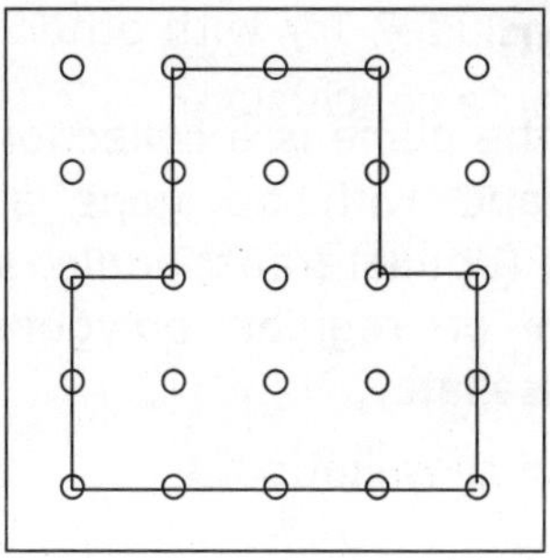

(b)

Take the following square as one unit.

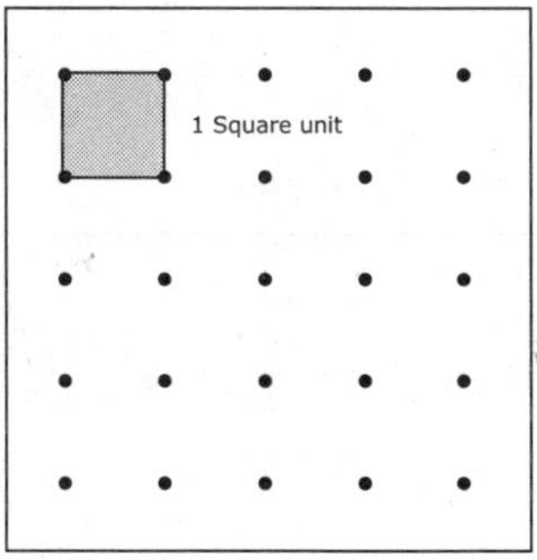

(c)

Then we can find twelve such small squares in the given shape. Thus area of the given shape is 12 square units.

Make the following shapes on the geoboard and calculate their area.

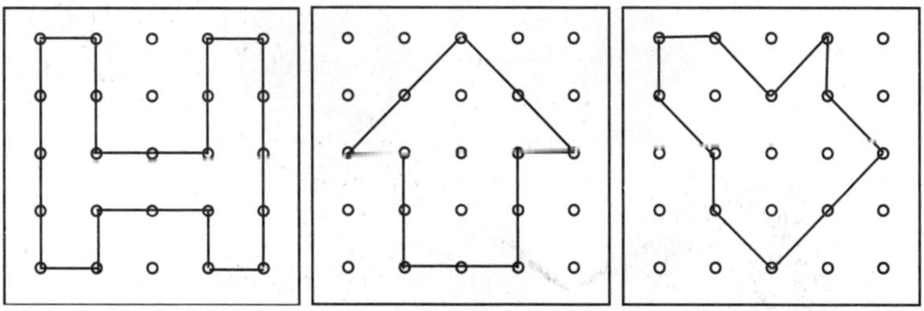

(d) *(e)* *(f)*

Fig.3.10

h) Tessellation

A tessellation of the plane is a collection of plane figures that fills the plane with no gaps and no overlaps. Regular polygons (equilateral triangle, square, hexagon) and combination of regular polygons tessellate. All quadrilaterals tessellate.

Some examples of tessellation:

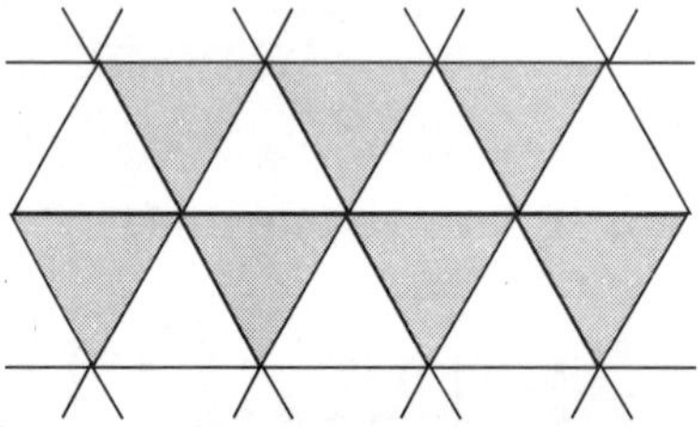

(a)

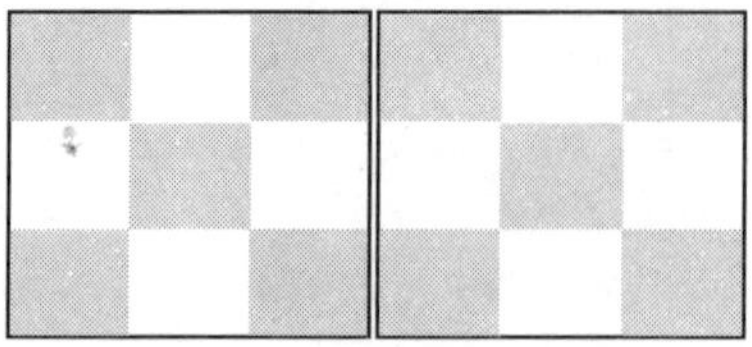

(b)

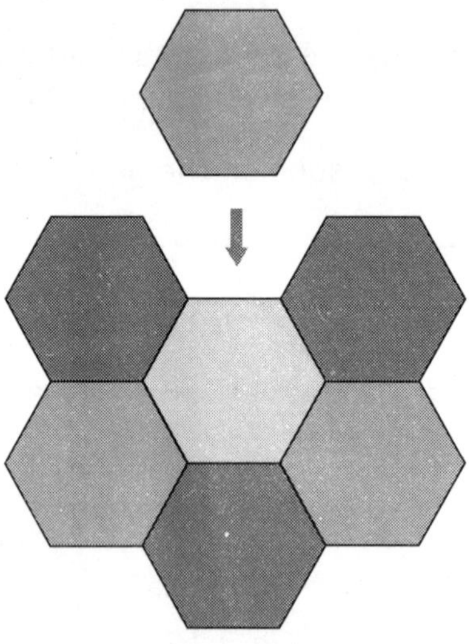

(c)

Materials required:

Tracing paper, pencil, thin cardboard, glue, a pair of scissors

Method:

1. Which of the following shapes tessellate? Trace or draw them on a paper. Cut along the boundary.
2. Now paste them on a thin cardboard and try to tessellate them(Cut several pieces of the same shape).

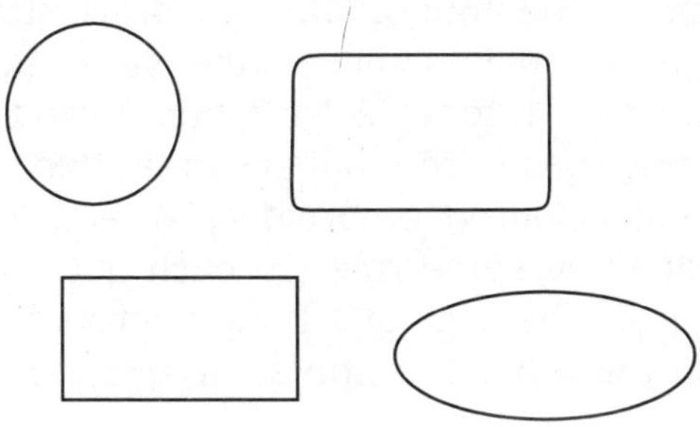

(d)

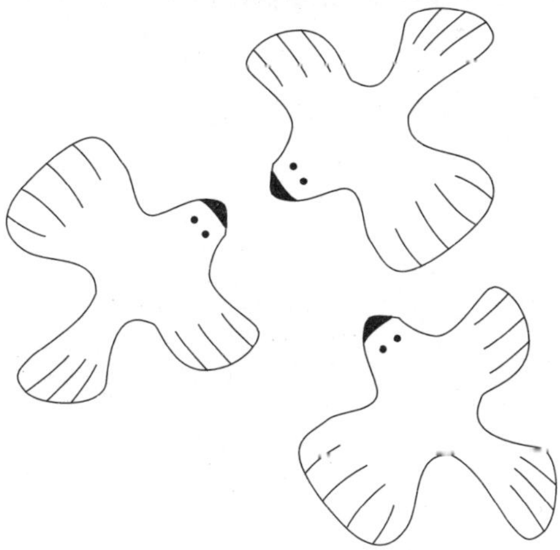

(e)

Fig. 3.11

i) String Art

Materials required:

Paper, pencil, ruler, compass, thin cardboard, needle, embroidery thread

Method:

Case I

1. Photocopy or draw the following graph axis.
2. In each of the four quadrants draw straight lines between two points having the same number. For example join 1 to 1, 2 to 2 etc. To make it more interesting paste the paper on a thin cardboard. Put some coloured embroidery thread in a needle and follow the same rule (In each quadrant put the needle from the back of 1 take it out and put it down through the other 1 in the same quadrant. Continue the process).
3. You will get an amazing pattern.

(a)

Example:

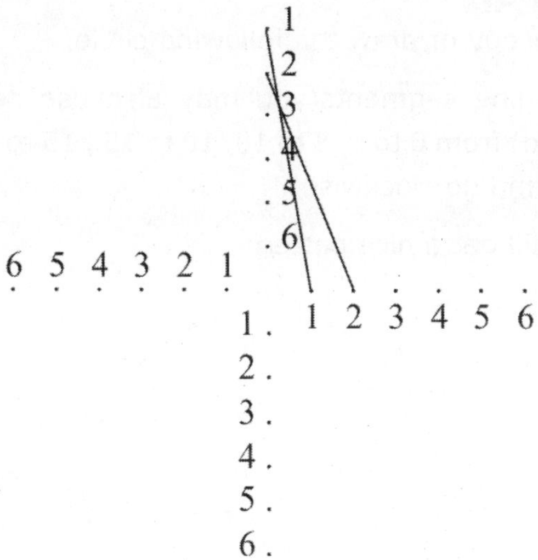

(b)

Fig. 3.12

Case II

1. Draw or photocopy the following square.
2. Draw straight lines between the four points having the same number. (You may use needle and thread if you like.) For example join 1 on top of the square with 1 on the remaining three sides of the square.

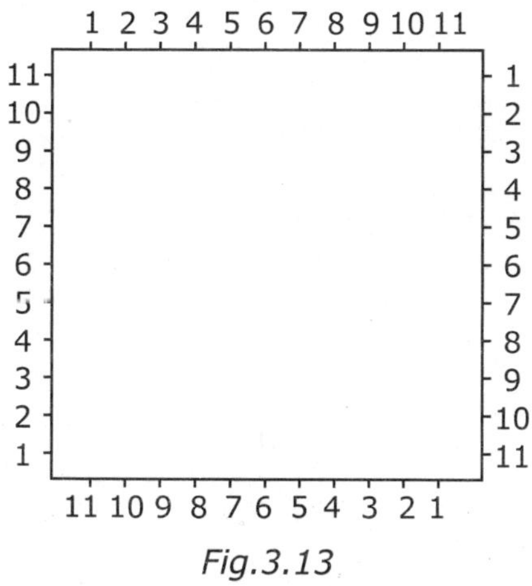

Fig.3.13

Case III

1. Photocopy or draw the following circle.
2. Draw line segments(you may also use needle and thread) from 0 to 5, 5 to 10, 10 to 15 , 15 to 4 etc(skip by 5 and go clockwise).
3. You will get a nice pattern.

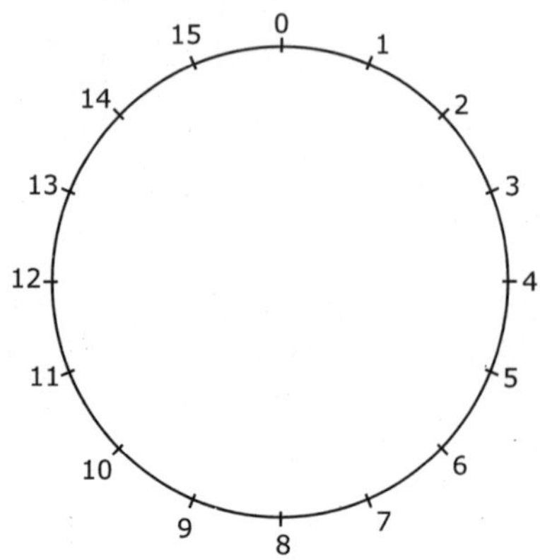

(a)

Example:

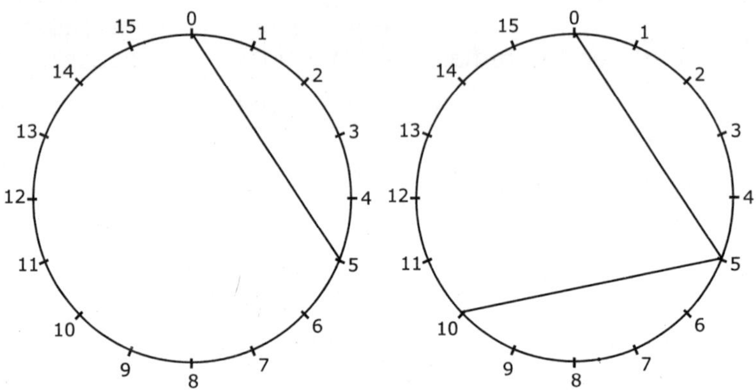

(b)

Fig. 3.14

j) Birth Day

Materials required:

Paper and pencil

Use the following charts to calculate the birth day(of the week) from the date of birth.

Month	Magic Number	Month	Magic Number
January	1*	July	0
February	4*	August	3
March	4	September	6
April	0	October	1
May	2	November	4
June	5	December	6

Table 3.3

* For leap year's use 0 for January and 3 for February.

Days of the Week	Magic Number
Sunday	1
Monday	2
Tuesday	3
Wednesday	4
Thursday	5
Friday	6
Saturday	0

Table 3.4

Method:

1. Write the last two digits of your birth year.

2. Divide the number by four(Ignore the remainder).
3. Find out the magic number of your birth month from the table given above.
4. Write the date on which you were born.
5. Add the numbers obtained in step 1 to step 4.
6. Divide this number by seven. The remainder will give you the day on which you were born(Check the day from the second table).

Example:

Suppose you were born on 14th February, 1999 then you would calculate the birth day as follows:

1. 99 (Last two digits of the birth year)
2. 99 ÷ 4 = 24 Remainder 3
3. The magic number of February is 4(Since '99 was not a leap year).
4. 14 (birth date)
5. 99 + 24 + 4 + 14 = 141.
6. 141 ÷ 7 = 20 Remainder 1
7. 1 corresponds to Sunday so the birth day of the person born on 14th February'99 was Sunday.

k) Napier Bones

Napier's bones were discovered by Napier in 1617. He used ten number bones (vertical strips) to multiply any number by a single digit number. The eleventh bone was used to represent the multiplier.

Materials required:

Paper, pencil and ruler

1	0	1	2	3	4	5	6	7	8	9
2	0/0	0/2	0/4	0/6	0/8	1/0	1/2	1/4	1/6	1/8
3	0/0	0/3	0/6	0/9	1/2	1/5	1/8	2/1	2/4	2/7
4	0/0	0/4	0/8	1/2	1/6	2/0	2/4	2/8	3/2	3/6
5	0/0	0/5	1/0	1/5	2/0	2/5	3/0	3/5	4/0	4/5
6	0/0	0/6	1/2	1/8	2/4	3/0	3/6	4/2	4/8	5/4
7	0/0	0/7	1/4	2/1	2/8	3/5	4/2	4/9	5/6	6/3
8	0/0	0/8	1/6	2/4	3/2	4/0	4/8	5/6	6/4	7/2
9	0/0	0/9	1/8	2/7	3/6	4/5	5/4	6/3	7/2	8/1

(a)

1. Photocopy the above chart (you may copy it) and cut out each number bone(vertical strip).
2. To multiply a four digit number pick up the bones of the respective numbers and place it next to the multiplier bone.
3. The multiplication proceeds from right to left starting with the rightmost bone in the row of the multiplier.
4. The rightmost number is written as it is.
5. The numbers within the parallelogram of the multiplier row is added with appropriate carry over as needed.

Example:

1. To multiply 9568 by 8 pick up the bones of 9, 5, 6 and 8.
2. Start from the rightmost of row 8.
3. Ones digit of the answer is 4.
4. Tens digit is 4(Add the numbers in the parallelogram 6 + 8 = 14).

5. Hundreds digit is 5(4 + 0 + 1 carry over from tens place).
6. Thousands digit is 6(4+ 2).
7. Ten thousands digit is 7.
8. Thus 9568 × 8 = 76544.

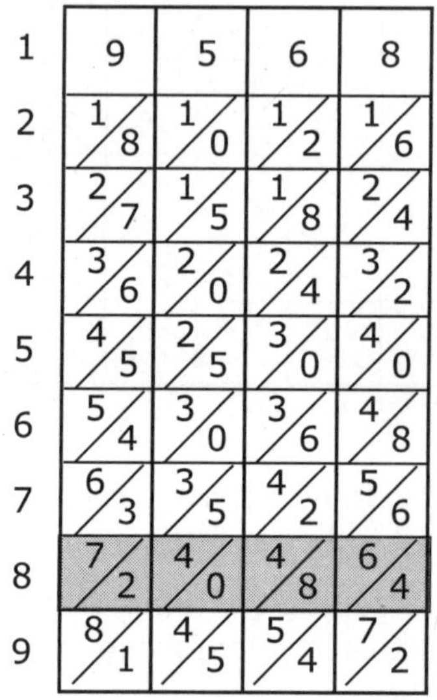

(b)

Fig. 3.15

Now try any multiplication using Napier bones.

I) Golden Rectangle

Two quantities are in Goden ratio(or Golden mean) if the ratio of the sum of two quantities(a and b) to the bigger quantity (a) is equal to the ratio of the bigger quantity(a) to the smaller quantity (b). It is generally denoted by phi(Φ).

Algebraically it can be represented as

$$\frac{a+b}{a} = \frac{a}{b} = \Phi$$

The above ratio has a unique solution given by

$$\Phi = \frac{1+\sqrt{5}}{2} = 1.6180339887...$$

Golden rectangle is a rectangle in which the ratio of the width to the height is the Golden ratio (1.618).

Materials required:

Paper, pencil, compass

Method:

1. Draw a square of any dimension.
2. Draw a dotted line at the middle of the square dividing the square into two equal rectangles. Extend the base of the square to the right.
3. Draw an arc of a circle from M with radius equal to MP. Let the arc meet the extended base at R.
4. Complete the Golden rectangle PQRS by taking QR as the width and QP as the height.

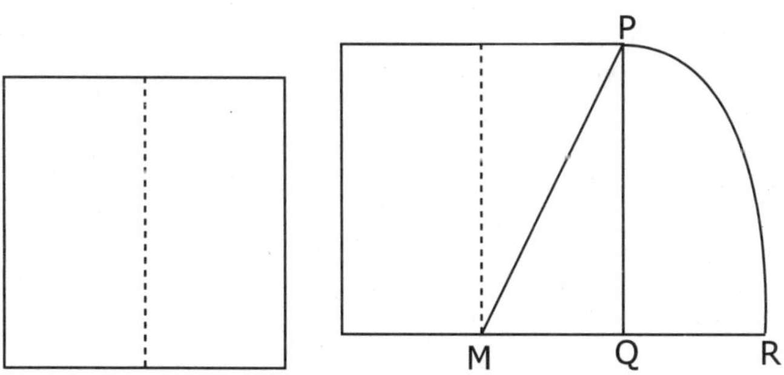

(a) (b)

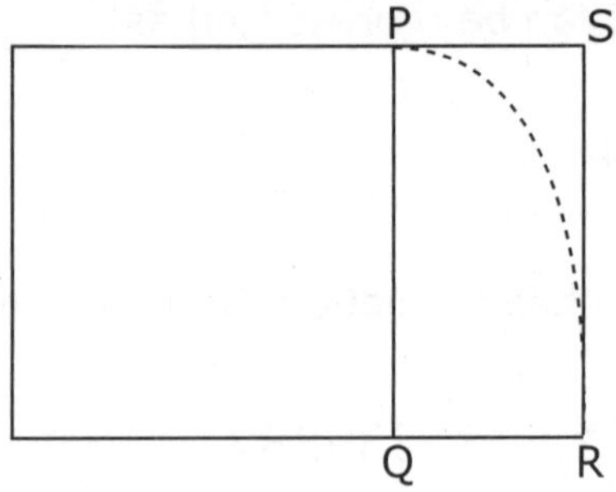

(c)

The Golden Spiral

1. A spiral can be drawn from the Golden rectangle above by dividing QRSP into squares and new Golden rectangles.
2. Keep the compass point at P and draw an arc from T to Q.
3. Next keep the compass point at V (QR = QV = RW) and draw an arc from Q to W.
4. Continue the process until you get the complete spiral.

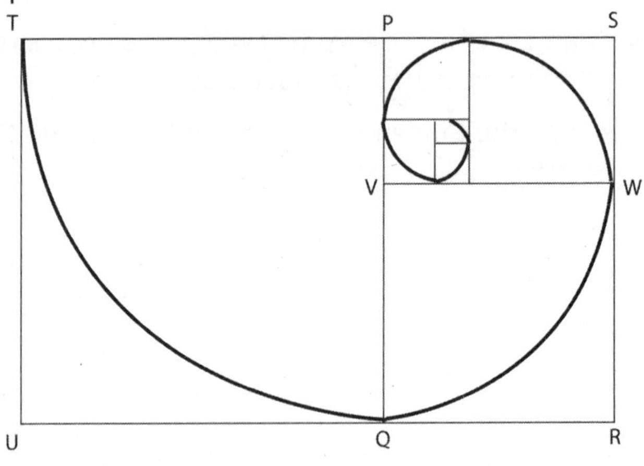

(d)

Fig. 3.16

+++

4

Trick and Quick

a) Squaring a two digit number beginning with 5.

1. To square a two digit number beginning with 5 add 25 to the digit at the ones place to get the first part of the answer.
2. Square the digit at the ones place to get the second part of the answer.

Example:

1. To find the square of 59 add 25 and 9 (25 + 9 = 34). Thus 34 is the first part of the answer.
2. Now $9^2 = 81$which is the second part of the answer.
3. Thus, $59^2 = 3481$.

b) Squaring Any Two Digit Number

1. To square any two digit number first replace the digit at the ones place by zero.
2. Next square the new two digit number.
3. Now square the digit at the ones place of the original number.
4. Finally multiply the digits at the tens place, ones place (of the original number) and 20.
5. Add all the answers.

Example:

1. To find the square of 78 first replace the digit at the ones place by zero. Thus replacing 8 by zero we get 70.
2. Next square the new two digit number. ($70^2 = 4900$)
3. Now square the digit at the ones place of the original number. ($8^2 = 64$)
4. Multiply the digits at the tens place, ones place and 20.($7 \times 8 \times 20 = 1120$)
5. Add all the answers ($4900 + 64 + 1120 = 6084$). Hence $78^2 = 6084$.

c) Squaring a Number Near 100.

Case I

1. If the number is above 100 and less than 110 add the number at the ones place to the three digit number. This gives the first part of the answer.
2. Square the digit at the ones place to get the second part of the answer.

Example:

1. To find the square of 107 add 107 and 7($107 + 7 = 114$). Thus 114 is the first part of the answer.
2. The second part of the answer is obtained by squaring 7. ($7^2 = 49$)
3. Hence $107^2 = 11449$.

Case II

1. If the number is above 110 add the last two digits to the three digit number. This gives the first part of the answer.
2. Square the last two digits and add as shown in the example below.

Example:

1. To find the square of 117 first add 117 and 17(117 +17 = 134).

2. Next find the square of 17(17^2 = 289).

   289
   134 (Add 2 and 4)

3. Thus, 117^2 = 13689

Case III

1. If the number is below 100 subtract the number from 100.

2. Next subtract the answer obtained in the previous step from the number. This forms the first part of the final answer.

3. Square the answer obtained in step 1. This is the last part of the answer.

Example:

1. To find the square of 93 first subtract 93 from 100(100 – 93 = 7)

2. Next subtract 7 from 93(93 – 7 = 86). Thus 86 is the first part of the answer.

3. Square 7 (7^2 = 49). Thus 49 is the last part of the answer.

4. Hence, 93^2 = 8649.

Exercise:

Using the short cut method find the square of the following numbers:

1) 57 2) 83 3) 109 4) 132 5) 94

d) The 7-11-13 Trick

1. Ask a friend to write any three digit number on a sheet of paper.
2. Ask him/her to multiply the number by 7, then by 11 and finally by 13. (You may give him/her a calculator if he/she wants.)
3. You will still get the answer quicker than your friend.

Trick:

Just write the three digit number twice to get the answer.

e) Calculation Race

You can have a calculation race with your friend.

Case I

1. Ask him/her to write any two digit number and multiply it manually by 3367.
2. You will always be able to get the answer before him/her without a calculator.

Example:

If your friend chooses 53 write 535353 and divide it by 3. Thus $535353 \div 3 = 178451$.

$(53 \times 3367 = 178451)$

Trick:

Write the two digit number thrice and divide it by 3.

Case II

1. Ask your friend to write a multiplication problem of multiplying a two digit number by another two digit number (except 99).
2. You write another similar problem.

3. Then challenge your friend to find the two products and add the two answers (without a calculator).
4. You will always be able to do it first.

Example:

1. Suppose your friend writes the problem as 79 × 34. Then you write the problem as 79 × 65.
2. Now 79 × 34 = 2686 and 79 × 65 = 5135.

   2686 + 5135 = 7821.

Trick:

1. Your multiplicand (the first number) should be the same as your friend's number (79 in this case).
2. Write the multiplier by subtracting it from 99 (99 – 34 = 65).
3. The first part of the answer is obtained by subtracting 1 from the multiplicand (79 -1 = 78). The last part of the answer is obtained by subtracting the multiplicand from 100(100 – 79 = 21).

f) Mnemonic

A mnemonic is a word, rhyme or any sentence which helps us to remember some rule, formula etc.

1. The following mnemonic help us in remembering the value of π (pi) to various decimal places. The number of letters in each word represents the digit.

   i. How I wish I could calculate pi.

   3.1 4 1 5 9 2

   ii. Can I tell a story regarding my family lives and views?

   3.1 4 1 5 9 2 6 5 3 5

   iii. Yes, I like a flute. Classical is lovely music. The music revolves immensely through knowledge.

   3.1 4 1 5 9 2 6 5 3 5 8 9 7 9

The following mnemonic helps us in remembering the trigonometric ratios:

Some People Have

Curly Brown Hair

Turned Permanently Black

* Use the first letter of each word to remember the ratios.

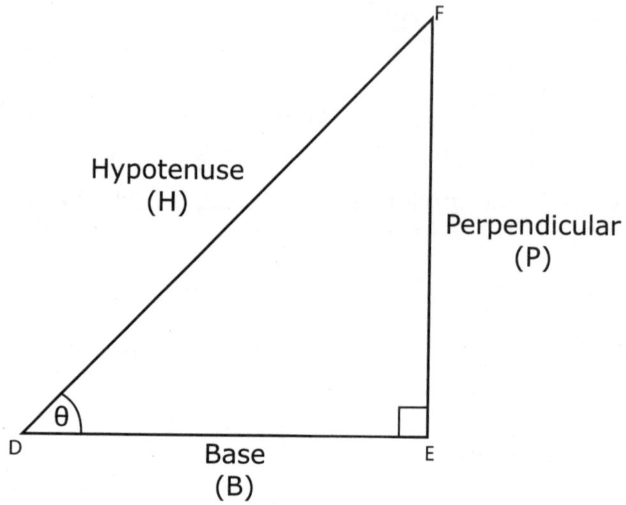

Fig.4.1

$$Sin\theta = \frac{Perpendicular(P)}{Hypotenuse(H)} \quad Cos\theta = \frac{Base(B)}{Hypotenuse(H)}$$

$$Tan\theta = \frac{Perpendicular(P)}{Base(B)}$$

g) Multiplying Two Digit Numbers by 11

1. To multiply a two digit number by 11 write the ones digit of the two digit number at the ones place of the answer.
2. Add the ones and tens digit to get the middle number of the answer.

3. Write the tens digit of the number at the hundreds place of the answer.

Examples:

i. To multiply 43 by 11 write 3 at the ones place of the answer.

   Add the ones and tens digit (4+3=7), 7 is the middle number of the answer.

   The tens digit 4 is in the hundreds place of the answer.

   Thus, 43 × 11 = 473

ii. To multiply 98 by 11 add the carry over to the number at hundreds place.

   17 (9 +8 = 17)

   9 8

   98 × 11 = 1078 (Add 9 and 1)

Exercise:

Using the above method find the following products:

1) 29 × 11   2) 88 × 11   3) 47 × 11

h) Multiplying Big Numbers by 11

1. To multiply big numbers by 11 first multiply the number by 10 then add the original number to the product.

Example:

To multiply 9876 by 11 first multiply it by 10(9876 × 10 = 98760)

Add 9876 to 98760(9876 + 98760 = 108636)

Exercise:

Using the above method find the following products:

1) 3456 × 11 2)12040 × 11 3) 8689 × 11

i) Some Multiplication Rules

1. Multiplication by 9: Multiply by 10 and subtract the original number.

Example:

To multiply 139 by 9 first multiply 139 by 10(139 × 10 = 1390). Next subtract 139 from 1390

(1390 - 139 = 1251) Thus 139 × 9 = 1251.

2. Multiplication by 13: Multiply by 3 and add 10 times the original number.

Example:

To multiply 253 by 13 first multiply 253 by 3(253 × 3 = 759). Next add 10 times the original number (759 + 253 × 10 = 759 + 2530 = 3289). Thus 253 × 13 = 3289.

3. Multiplication by 15: Multiply by 10 and add 5 times the original number.

Example:

To multiply 76 by 15 first multiply 76 by 10 (76 × 10 = 760). Next add 5 times the original number (760 + 5 × 76 = 760 + 380 = 1140). Thus 76 × 15 = 1140.

4. Multiplication by 17: Multiply by 7 and add 10 times the original number.

Example:

To multiply 321 by 17 first multiply 321 by 7(321 × 7 = 2247). Next add 10 times the original number (2247 + 10 × 321 = 2247 + 3210 = 5457). Thus 321 × 17 = 5457.

5. Multiplication by 19: Multiply by 20 and subtract the original number.

Example:

To multiply 431 by 19 first multiply 431 by 20(431 × 20 = 8620). Next subtract the original number(8620 - 431 = 8189).

6. Multiplication by 99: Multiply by 100 and subtract the original number.

Example:

To multiply 89 by 99 first multiply by 100 (89× 100 = 8900). Next subtract the original number (8900 - 89 = 8811).

Exercise:

Using the above rules find the following products:

1) 58 × 9 2) 156 × 13 3) 44 × 15 4) 49 × 17
5) 268 × 19 6) 75 × 99

j) Time Constant

1. This method gives a simple way of adding or subtracting units of time(hours and minutes) as compared to the conventional method.
2. To add two time just remove the hours and minutes and do normal addition.
3. Next add time constant 40 to the answer.
4. Finally write minutes for the last two digits and hours for the remaining digits.
5. The same method can be applied for subtracting time. Instead of adding 40 subtract 40.

Example:

1. To add 5 hours 45 minutes and 7 hours 35 minutes just remove hours and minutes and add 545 and 735 (545 + 735 = 1280).
2. Next add 40 to the answer(1280 + 40 = 1320).
3. Thus 5 hours 45 minutes + 7hours 35 minutes = 13hours 20 minutes.

Exercise:

Using the above rule solve the following:

1. 7 hours 55 minutes + 8 hours 25 minutes
2. 11 hours 35 minutes – 4 hours 45 minutes

k) Easy Division by 5

1. To divide a big number by 5 just multiply it by 2 and put the decimal point after one place from the right.

Example:

To divide 3987 by 5 just do 3987 × 2 = 7974 and shift the decimal i.e. 797.4

Exercise:

Using the above rule divide the following:

1) 4389 ÷ 5 2) 11567 ÷ 5 3) 70953 ÷ 5 4) 88997 ÷ 5

l) Easy Division by 9

1. A repeated pattern is produced when a number smaller than 9 is divided by 9.

Example:

1. 7 ÷ 9 = 0.777... (This will work if the dividend is from 1 to 8 and the divisor is 9)
2. 86 ÷ 99 = 0.8686... (This will work if the dividend is from 10 to 98 and the divisor is 99)

3. 5 ÷ 99 = 0.0505... (This will work for dividends 1 to 9 and divisor 99)
4. 594 ÷ 999 = 0.594594 ...(This will work for dividenas 100 to 998 and divisor 999)
5. 74 ÷ 999 = 0.074074 ... (This will work for dividends 10 to 99 and divisor 999)
6. 3 ÷ 999 = 0.003003... (This will work for dividends 1 to 9 and divisor 999)

Exercise:

Using the above rule divide the following :

1) 8 ÷ 9 2) 4 ÷ 99 3) 497 ÷ 999 4) 67 ÷ 999
5) 8 ÷ 999

m) Perfect Number

1. A perfect number is a number which is equal to the sum of its factors (including 1 but excluding itself. The smallest perfect number is 6 = 1+ 2 + 3.
2. To find other perfect numbers use the formula: $2^{n \quad 1} \times (2^n - 1)$ where n is a prime number.

Example:

Perfect number after 6 is $2^{3-1} \times (2^3 - 1) = 2^2 . 7 = 4 \times 7 = 28$.

Exercise:

Write three perfect numbers after 28.

n) Checking four Operations

Addition:

1. Add the digits of both the addends separately till you reach a one digit number.
2. Do the same for the answer(sum).

3. Now add the single digits of the addends. This should be equal to the single digit of the sum if the addition is correct.

Example:

2978	Addend	(2 + 9 + 7 + 8 = 26, 2 + 6 =8)
+ 5096	Addend	(5 + 0 + 9 + 6 = 20, 2 + 0 = 2)
8074	Sum	(8 + 0 + 7 + 4 = 19, 1 + 9 = 10)

8 + 2 = 10 same as the single digit obtained from the sum. Hence the addition is correct.

Subtraction:

Add the difference to the subtrahend (the number being subtracted). If it gives the minuend (the number from which it is subtracted) the subtraction is correct.

Example:

9687	Minuend
- 2899	Subtrahend
6788	Difference

Checking:

6788	Difference
+ 2899	Subtrahend
9687	Minuend

Multiplication:

1. Add the digits of both the multiplicand and the multiplier separately till you reach a one digit number.
2. Do the same for the product.
3. Multiply the one digit number of the multiplicand and the multiplier. If this answer is equal to the one digit number obtained from the product the answer is correct.

Example:

245	Multiplicand	(2 + 4 + 5 = 11, 1 + 1 = 2)
× 927	Multiplier	(9 + 2 + 7 = 18, 1 + 8 = 9)
227115	Product	(2 + 2 + 7 + 1 + 1 + 5 = 18, 1 + 8 = 9)

Multiplying the one digit numbers of the multiplicand and the multiplier we get 2 × 9 = 18, 1 + 8 = 9 which is equal to the one digit number of the product. Hence the multiplication is correct.

Division (without remainder):

1. We can check division (without remainder) by converting the division problem into a multiplication problem (since division is inverse of multiplication) and using the same rule of multiplication.

Example:

4704 ÷ 48 = 98 can be written as

```
   98 (9 + 8 = 17, 7 + 1 = 8)
 x 48 (4 + 8 = 12, 1 + 2 = 3)
 4704 (4 + 7 + 0 + 4 = 15, 1 + 5 = 6)
```

8 × 3 = 24, 2 + 4 = 6 which is equal to the one digit number of the product. Hence the division is correct.

Division (with remainder):

To check division (with remainder) is slightly cumbersome. We can check it by using the following formula:

Dividend = Divisor × Quotient + Remainder

Example:

5309 (Dividend) ÷ 29(Divisor) = 183(Quotient) 2 (Remainder)

183 × 29 = 5307, 5307 + 2 = 5309

Thus, 5309 = 183 × 29 + 2

o) Quick Division by 9

Case I (When the dividend is a two digit number)

1. When you divide a two digit number by 9 the first digit gives the quotient and the sum of the digits gives you the remainder.
2. If remainder is more than 9 we divide the remainder again by 9 and add that quotient to the previous quotient.

Example:

1. 53 ÷ 9 = 5 (First digit of 53), Remainder 8(5 + 3)
2. 93 ÷ 9 = 9 Remainder 12 but remainder cannot be greater than the divisor hence 12 ÷ 9 = 1 Remainder 3. Thus 93 ÷ 9 = 9 + 1 = 10 Remainder 3.

Case II (When the dividend has any number of digits)

1. Write the first number as it is.
2. Next add the first number to the second, sum to the third, new sum to the fourth etc.
3. When you add the last number of the dividend the new sum is the remainder if it is less than 9. Otherwise divide it once again by 9 using the case I rule and add the quotient to the last digit(8 in this case)

Example:

1. To divide 321023 by 9 keep the first digit as it is i.e. 3.
2. Now add 3 to 2 which equals 5, 5 to 1 which gives 6, 6 to 0 which gives 6, 6 to 2 which gives 8 and finally 8 to 3(last number of the dividend) which gives 11.
3. Now 11 is greater than 9 hence using case I rule, 11 ÷ 9 = 1 remainder 2. Add this 1 to the last digit of the quotient(8 + 1 = 9).

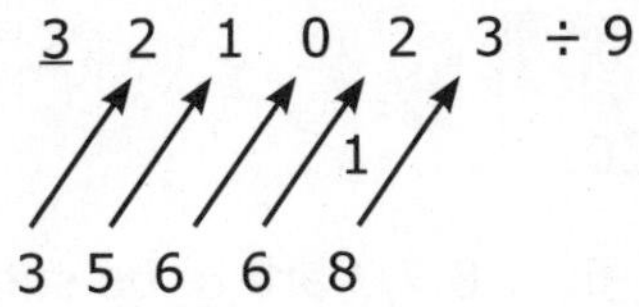

8 + 3 = 11, 11 ÷ 9 = 1 Remainder 2

4. Thus the quotient is 35669 and the remainder is 2.

Exercise:

Using the above rule divide the following:

1) 57 ÷ 9   2) 96 ÷ 9   3) 4587 ÷ 9   4)10897 ÷ 9

p) Vedic Mathematics

'Vedic Mathematics' is the name given to the ancient system of mathematics that was rediscovered by Sri Bharati Krsna Tirthaji Maharaj between 1911 and 1918. He was an Indian scholar well versed in Sanskrit, Mathematics, Astronomy etc. The roots of Vedic Mathematics are found in Vedas written hundreds of years ago. The system is based on 16 short Sanskrit verses known as "Sutras" which are actually word formulae describing easy ways of solving mathematical problems. Some of the sutras are given below.

i) Nikhilam Navatashcaramam Dashatah (All from 9 and the last from 10)

- This is a method used to subtract or multiply numbers which are close to an *Aadhara* (base) such as 10, 100, 1000, 10000 etc.

Subtraction:

1. All numbers are subtracted from 9 except the last(ones digit) which is subtracted from 10.

Example:

```
  1 0 0 0 0
-   8 7 3 6   ( 8,7,3 are to be subtracted from 9
              and last number 6 from 10)
    1 2 6 4
```

Multiplication (when both the numbers are lower than the base):

1. When both the numbers are lower than the base first write the deviation of the two numbers from the base. The number of digits of the deviation should be same as the number of zeros of the base. For example, if the base is 100 and the deviation is 9 it should be written as 09 but if the base number is 1000 and the deviation is 9 it should be written as 009.
2. To get the last part of the answer multiply the two deviations. The product should have the same number of digits as the number of zeros of the base. For example, if the base is 100 the product should have two digits. In case the product has three digits the number in the hundreds place is added to the ones digit of the first part of the answer (See example 2 of multiplication (when both the numbers are higher than the base).
3. Next subtract the first number and the second deviation or the second number and the first deviation to get the first part of the answer.

Example 1:

1. To find the product of 92 by 93 we will take 100 as the base and find the deviations.
2. To get the last part of the answer multiply the deviations ($8 \times 7 = 56$).

3. Next subtract the first number from the second deviation or vice versa (92 – 7 = 85 or 93 – 8 = 85).

Step1.

```
  92  08 |  (100 - 92 = 08)
x 93  07 ↓  (100 - 93 = 07)
---------
      56
```

Step 2.

```
  92  08 |
x 93  07 ↓
---------
  85  56
   |
   ↓
 92 - 7
    or
 93 - 8
```

Example 2:

See the following example carefully.

Step1.

```
  98 02 |  (100 - 98 = 02)
x 96 04 ↓  (100 - 96 = 04)
--------
  94 08
   |
   ↓
(98 - 4 = 94)
      or
(96 - 2 = 94)
```

In the above example, 948 is not the correct answer since the deviations are not written properly.

Step2.

```
   98  2  |   (100 - 98 = 02)
x  96  4  ↓   (100 - 98 = 04)
----------
   94  8
   ↓
(98 - 4 = 94)
      or
(96 - 2 = 94)
```

Multiplication (when both the numbers are higher than the base):

1. The first two steps are same as above.
2. In the third step instead of subtracting we add the first number with the second deviation or the second number with the first deviation.

Example 1:

Step1.

```
   108  08  |   (108 - 100 = 08)
x  103  03  ↓   (103 - 100 = 03)
------------
        24
```

Step 2.

```
   108  08  |
x  103  03  ↓
------------
   111  24
   ↓
108 + 3
   or
103 + 8
```

Example 2:

Step1.

```
  112  12   | (112 - 100 = 12)
x 113  13   ↓ (113 - 100 = 13)
-----------
  125 156
   ↓
112 + 13
   or
113 + 12
```

In the above example, since the base is 100 the last part of the answer should have two digits but in this case it has three digits. Hence the digit 1(the digit at the hundreds place) is added to 5 (the ones digit of the first part of the answer). Thus, the final answer is

```
  112  12   | (112 - 100 = 12)
x 113  13   ↓ (113 - 100 = 13)
-----------
  126  56
   ↓
 125 + 1
```

ii) Ekadhikena purvena (By one more than the previous one)

Case I

1. To square a number ending with 5 always write 2 at the tens place and 5 at the ones place (i.e 25 in the end).
2. To write the first part of the answer multiply the number other than 5 (in the question) by its next consecutive number (one more than the previous number).

Example:

$65^2 = 4225$

42 25

↓ ↓

Always 25.

6 × 7(number other than 5 in the question is 6 and its next consecutive number is 7)

Case II (Dividing 1 by 19, 29, 39 ...)

1. To divide 1 by 19 start with the last digit as 1.
2. Multiply this with Ekadhikena i.e. one more than the digit at the tens place of the divisor. If the divisor is 29 multiply by 3, if the divisor is 39 multiply by 4. In this case it is 2 (the key digit) ⟶ 21
3. Keep multiplying the last digit by 2 taking care of the carry over ⟶ 8421(2 × 1 = 2, 2×2 = 4, 2×4=8)
4. Multiply 8 by 2 ⟶ 68421 carry 1.
5. Multiply 6 by 2 =12 + carry 1 = 13 (write 3 and again carry 1) ⟶ 368421
6. Continue the multiplication till you reach 10. Ignore the 1 and put a decimal point.
7. The final result is 0.052631578947368421.

iii) Urdhva Tiryagbhyam(Vertically and crosswise)

1. This sutra is used to find the product of numbers with easy calculations.
2. Using this sutra we first multiply vertically from the ones place taking care of carry over if any.
3. Next we multiply crosswise and add. We write the carry over (if any on top of the tens place).
4. Finally we again multiply vertically the numbers at the tens place and add the previous carry over.

Example:

(1)

4 2
x 2 3
6 $(2 \times 3 = 6)$

(2)

4 2
x 2 3
6 $(4 \times 3 + 2 \times 2 = 16)$

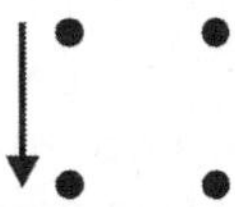

(3)

1
4 2
x 2 3
9 $(4 \times 2 + 1 = 9)$

Thus,

4 2
x 2 3
9 6 6

iv) Anurupye Sunyamanyat (If one is in ratio, the other one is zero))

1. This sutra is used to solve simultaneous linear equations in which the ratio of the coefficients of one variable is equal to the ratio of the independent terms.
2. In such a situation just equate the other variable to zero.

Example:

243x + 419y = 2145

721x + 1257y = 6435

1. The coefficients of y are in the ratio 1:3.
2. The ratio of the independent term is also 1:3 (2145:6435 = 1:3). Thus by the above sutra we can straight away equate x to zero.
3. Now substituting x = 0 in equation (1) we get 419y = 2145.

Thus, y = 2145/419.

v) *Sankalana Vyavakalanabhyam (By addition and by subtraction)*

1. This sutra is used to solve those simultaneous equations in which the coefficients of x and y are interchanged.
2. To solve such equations first add the two equations then subtract them. This simplifies the two equations.

Example:

$1955x - 476y = 2482$ (1)

$476x - 1955y = -4913$ (2)

Adding equation (1) and (2) we get,

$2431x - 2431y = -2431$

Or, $x - y = -1$(Dividing throughout by 2431)

Subtracting equation (1) and (2) we get,

$1479x + 1479y = 7395$

Or, $x + y = 5$ (Dividing throughout by 1479)

Now it's easy to solve $x - y = -1$ and $x + y = 5$. We get, $x = 2$ and $y = 3$.

Exercise:

A) Use *Nikhilam Navatashcaramam Dashatah sutra* to solve the following:

1) $1000 - 887$ 2) $100000 - 68975$ 3) 98×94
4) 95×97 5) 107×104 6) 109×103

B) Use *Ekadhikena purvena sutra* to solve the following:

1) 75^2 2) 85^2 3) $1 \div 29$ 4) $1 \div 39$

C) Use *Urdhva Tiryagbhyam sutra* to solve the following:

1) 46×83 2) 74×39 3) 68×53 4) 89×39

D) Use *Anurupye Sunyamanyat sutra* to solve the following equations:

i) $3x + 7y = 2$ ii) $24x + 31y = 48$

$4x + 21y = 6$ $96x + 60y = 192$

E) Use *Sankalana Vyavakalanabhyam sutra* to solve the following equations:

i) $45x - 23y = 113$ ii) $659y + 956z = 4186$

$23x - 45y = 91$ $956y + 659z = 3889$

+++

5

Puzzles

a) Compass Direction

What digit has the least frequent occurrence between the numbers 1 and 1,000?

To find the answer follow the compass directions:

1. Start from the star and go east three numbers.
2. Go south east one number.
3. Go south west two numbers.
4. Finally go west two numbers and south one number.

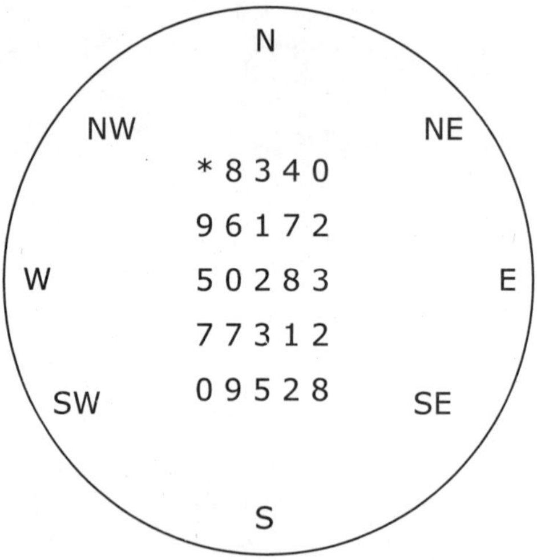

Fig 5.1

b) Proof of 1 = 2 using Algebra.

Proof:

Let $a = b = 1$

Then $a^2 = b^2$

Or, $a^2 = ab$ (since $b^2 = b \times b$ and $b = a$)

Or, $a^2 - b^2 = ab - b^2$ (subtracting b^2 from both sides)

Or, $(a + b)(a - b) = b(a - b)$

Dividing both sides by $(a - b)$ we get:

$$(a + b) = b$$

Substituting the values of a and b we get

$$1 + 1 = 1$$

Or, $2 = 1$

Where is the fallacy?

c) Proof of π (pi) = 3 using Algebra

Proof:

Let $x = \frac{\pi + 3}{2}$

Or, Let $2x = \pi + 3$

Multiplying both sides by $(\pi - 3)$ we get,

$$2x(\pi - 3) = (\pi + 3)(\pi - 3)$$

Or, $2\pi x - 6x = \pi^2 - 9$

Or, $\pi^2 - 2\pi x = 9 - 6x$

Adding x^2 to both sides we get,

$$\pi^2 - 2\pi x + x^2 = 9 - 6x + x^2$$

$(x - \pi)^2 = (x - 3)^2$

Taking square root on both sides we get,

$x - \pi = x - 3$

Or, $\pi = 3$ (Cancelling x from both sides)

But we know that π is an irrational number and its value is 3.141... So where is the fallacy?

d) Alphamagic Square

A magic square is a square of numbers in which every column, every row and both the diagonals add up to the same number. This number is often called the magic sum. In a 3 × 3 magic square the magic sum is found by multiplying the second row second number by 3. For example, in the following magic square the second row second number is 5. Hence, 5 × 3 = 15 which is the magic sum.

8	1	6
3	5	7
4	9	2

Table 5.1 (a)

A 3 × 3 alphamagic square is a magic square for which the number of letters in the word for each number generates another magic square. For example, if the alphamagic square has a number 18 in some position the corresponding magic square will have the number 8 in the same position. This is because the word eighteen has eight (8) letters in it. Similarly if the alphamagic square has five in some position the corresponding magic square will have 4 in the same position.

Complete the following alphamagic square and derive the other magic square.

		18
28	15	

Table 5.1 (b)

e) T Puzzle

Trace the following shapes on a paper. Cut along the outline and paste them on a thin cardboard. Now using the cardboard shapes form a "T". (Use all the shapes.)

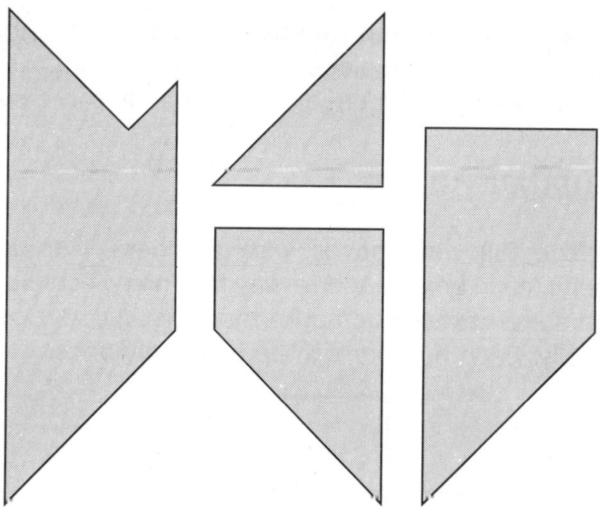

Fig.5.2

f) Doctor's Medicine

A doctor gave three tablets to Sunil and told him to take one every half an hour. How long will the tablets last?

g) Same Result

Which number when multiplied by 5/4 gives the same result as when added to 5/4?

h) Coloured Wrist Bands

Three basket ball players Mr. White, Mr. Blue and Mr. Red went to an ice cream parlour after their match. While eating the ice cream Mr. Blue said, "Did you notice that the colour of our wrist bands are different from our names?" The person who was wearing the red wrist band said, "Yes you are right! You have a good observation." Can you tell which player is wearing which colour wrist band?

i) Cut the Cake

Deepayan invited seven friends on his birthday. All eight of them wanted equal pieces of the round cake. How will his mother make eight equal pieces in only three cuts?

j) Number Puzzle

Fill in the following boxes with numbers 1 to 8 (one number in each box) in such a way that no two consecutive numbers are next to each other horizontally, vertically, or diagonally. Each number can be used only once.

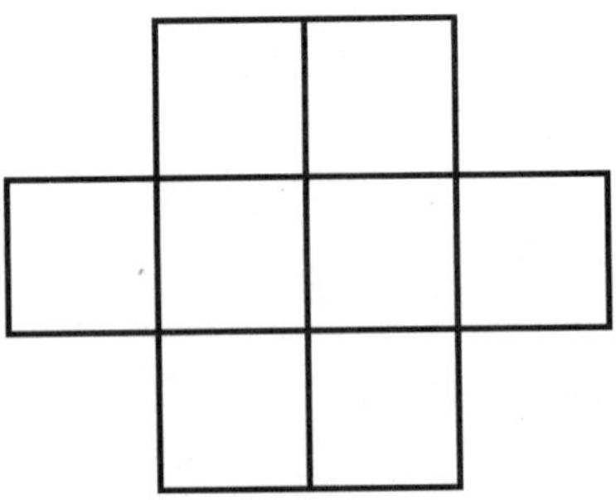

Fig. 5.3

k) Oldest Number Puzzle

As I was going to St. Ives,

I met a man with seven wives,

Every wife had seven sacks,

Every sack had seven cats,

Every cat had seven kits,

Kits, cats, sacks and wives,

How many were going to St.Ives?

l) Roman Numbers

In each of following equations move one straw to a new position to make the equations correct.

I – III = II

VI – IV = IX

m) Friar's Puzzle

Put ten counters or coins or buttons on the following grid in such a way that each row, each column and each diagonal has an even number of counters or coins or buttons.

Table 5.4

n) Car Speed

Mr. Nair drives a car at a uniform speed. He sees a milestone showing a two digit number. After travelling for one hour he sees another milestone with the same digits but in the reverse order. He keeps driving and another hour later he sees the third milestone containing the same two digits with a zero in between. What is his average speed?

o) Morse Code

Morse code is a system in which the English letters and the Hindu Arabic numbers are represented by DOTS and DASHES. (See the table) It was used over telegraph lines to send messages. During a war, spies often communicated using this code in order to keep it a secret. You can share your secret with your friend using Morse code.

Use the following Morse Code chart to find the only number in English language that has all its letters in alphabetical order.

··—· , — — — , ·—· , — , —·— —

(Comma separates each letter)

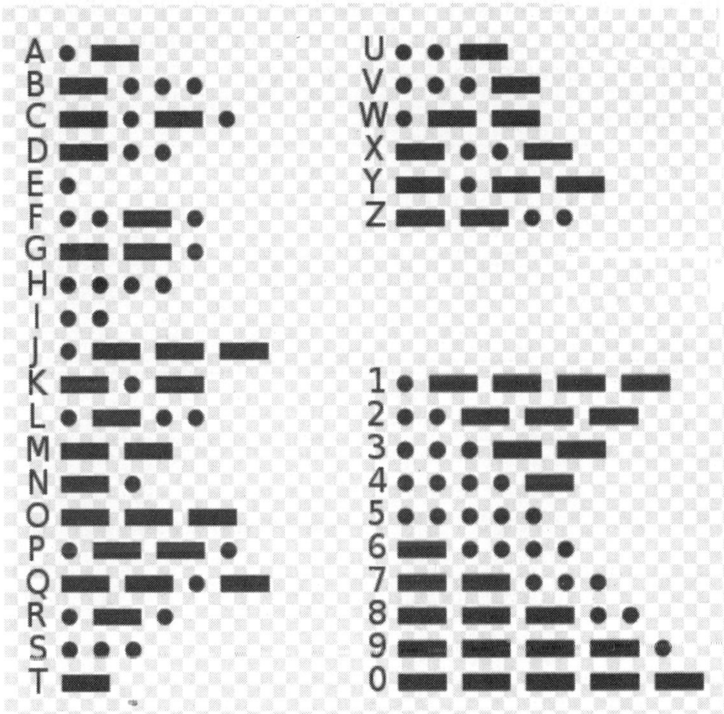

Fig. 5.5

p) Walking Time

Mr. Khan's driver always arrived at the train station at five o'clock sharp to pick him up and drive him home. One day, Mr. Khan arrived one hour early. He did not wait for the car and started walking home. He was eventually picked up by his driver on the way. He reached home 20 minutes earlier than usual. How long did he walk before he was picked up by his driver?

q) Number Fence

Make a fence in each of the following grids. The fence must be a closed loop that connects dots horizontally or vertically, but not diagonally. The numbers in the grid tells how many sides of the fence go around

Example:

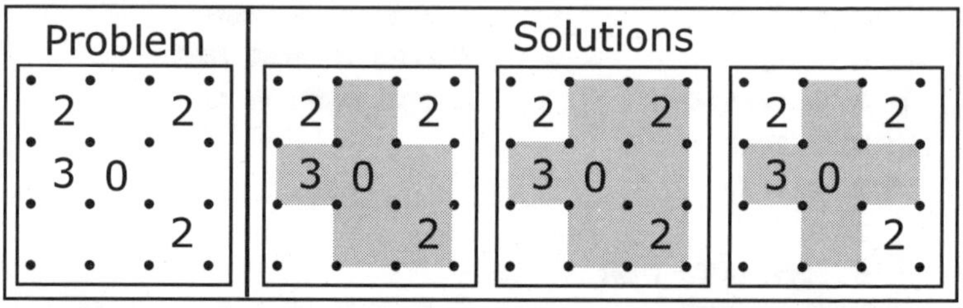

Fig. 5.6 (a)

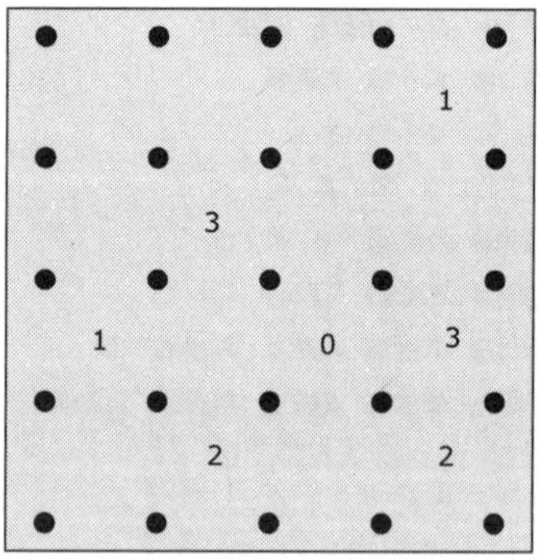

Fig.5.6(b)

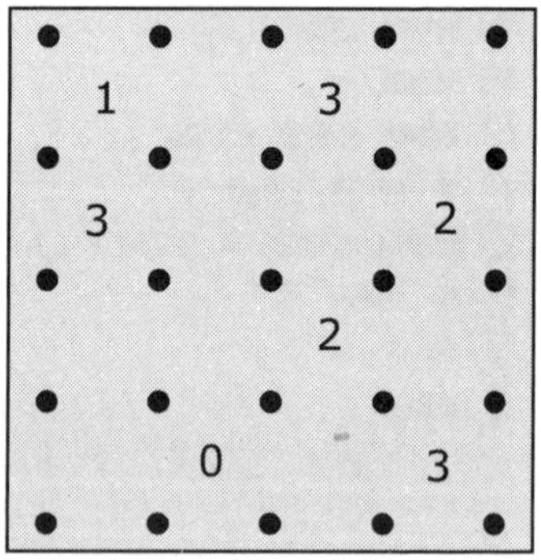

Fig.5.6(c)

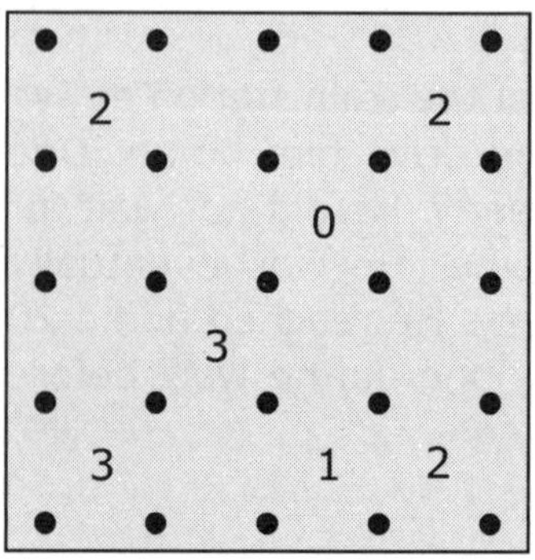

Fig.5.6 (d)

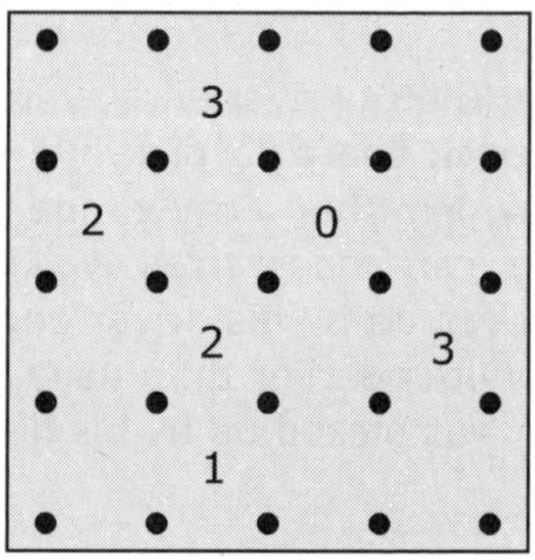

Fig.5.6(e)

r) Grandmother's Gift

Christopher's grandmother got a pair of shoes, a shirt and a pant for him as a Christmas present. However she was not sure of his size and wanted him to try them out. (She wanted to return the item to the shop in case the size was not correct.) Christopher tried them in his room came out and said thanks to his grandma. He handed a calculator to his grandmother and asked her to key in 35007 to see the first word of his reply and key in 3045 to see the second word. Can you find out his reply?

s) Burning Rope

There are two ropes made from some non-homogeneous material. Due to this special material, the rope takes exactly one hour to burn if lighted from one of the ends. However, as the material is not uniform throughout the rope, different parts of the rope may take different time to burn. For example, one half may take 20 minutes where as the other half may take 40 minutes to burn. How will you measure 45 minutes using these two ropes?(Assume there is no time lag in burning)

t) Age

Two mathematics teachers had their birthday on the same date. They wished each other. Then one of them said, "You will have such an age only after 11 years." The other teacher replied, "Yes, but you will have such an age only after 96 years." How old were the two teachers?

u) Tangram

Trace the following shape and cut along the lines. Paste the shapes on a thin cardboard to make the tangram puzzle set.

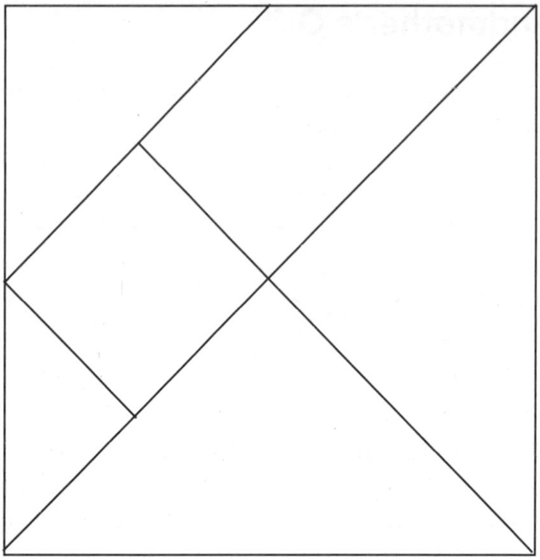

Fig. 5.7 (a)

Make the following shapes using the tangram puzzle set:

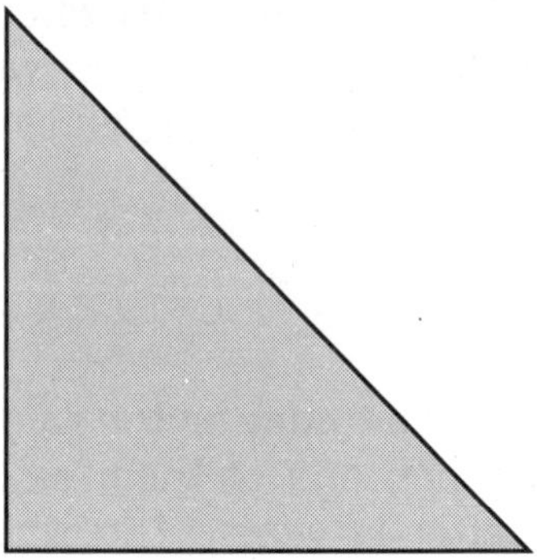

Fig. 5.7 (b)

Fig. 5.7 (c)

Fig. 5.7 (d)

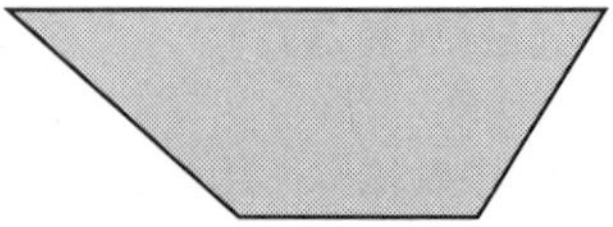

Fig. 5.7 (e)

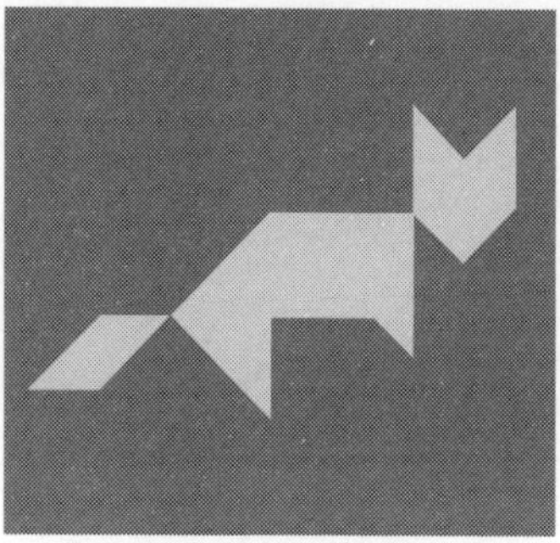

Fig. 5.7 (f)

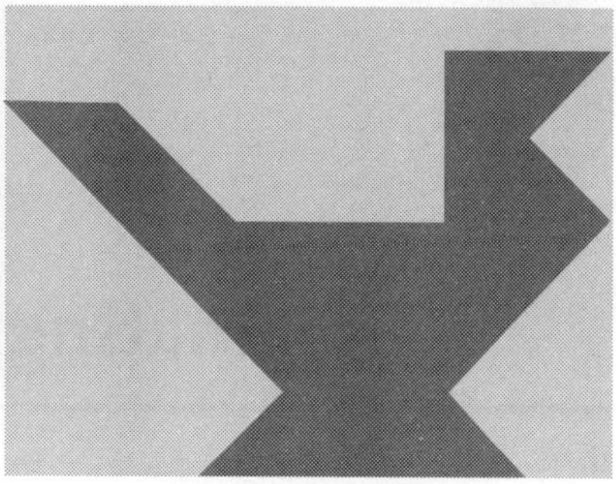

Fig. 5.7 (g)

Fig. 5.7 (h)

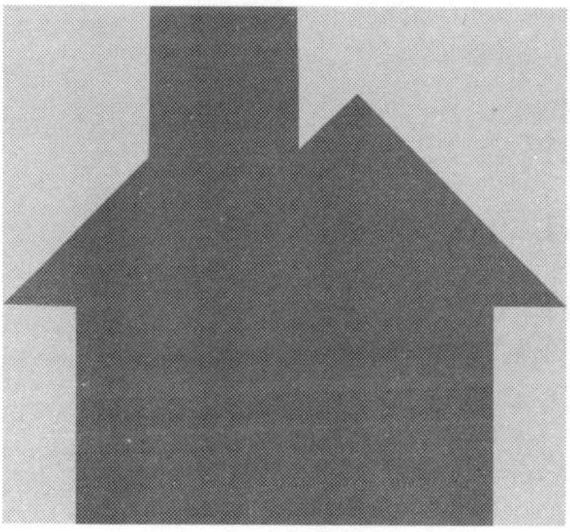

Fig. 5.7 (i)

6

Fascinating patterns

a) Number Patterns

$1^2 = 1$

$11^2 = 121$

$111^2 = 12321$

$1111^2 = 1234321$

$3^2 = 9$

$33^2 = 1089$

$333^2 = 110889$

$3333^2 = 11108889$

$67^2 = 4489$

$667^2 = 444889$

$6667^2 = 44448889$

$66667^2 = 4444488889$

Exercise:

Find the following squares by seeing the number pattern above.

1. 11111^2
2. 111111^2
3. 33333^2
4. 333333^2

ii)

$$1 \times 8 + 1 = 9$$
$$12 \times 8 + 2 = 98$$
$$123 \times 8 + 3 = 987$$
$$1234 \times 8 + 4 = 9876$$
$$12345 \times 8 + 5 = 98765$$
$$123456 \times 8 + 6 = 987654$$
$$1234567 \times 8 + 7 = 9876543$$
$$12345678 \times 8 + 8 = 98765432$$

iii)

$$0 \times 9 + 1 = 1$$
$$1 \times 9 + 2 = 11$$
$$12 \times 9 + 3 = 111$$
$$123 \times 9 + 4 = 1111$$
$$1234 \times 9 + 5 = 11111$$
$$12345 \times 9 + 6 = 111111$$
$$123456 \times 9 + 7 = 1111111$$
$$1234567 \times 9 + 8 = 11111111$$
$$12345678 \times 9 + 9 = 111111111$$

iv)

$$9 \times 9 + 7 = 88$$
$$98 \times 9 + 6 = 888$$
$$987 \times 9 + 5 = 8888$$
$$9876 \times 9 + 4 = 88888$$
$$98765 \times 9 + 3 = 888888$$
$$987654 \times 9 + 2 = 8888888$$
$$9876543 \times 9 + 1 = 88888888$$

v)

$$1 \times 9 + 1 + 9 = 19$$
$$2 \times 9 + 2 + 9 = 29$$
$$3 \times 9 + 3 + 9 = 39$$
$$4 \times 9 + 4 + 9 = 49$$
$$5 \times 9 + 5 + 9 = 59$$
$$6 \times 9 + 6 + 9 = 69$$
$$7 \times 9 + 7 + 9 = 79$$
$$8 \times 9 + 8 + 9 = 89$$

vi) 1259 × 2 + 1 = 2519

839 × 3 + 2 = 2519

629 × 4 + 3 = 2519

503 × 5 + 4 = 2519

419 × 6 + 5 = 2519

359 × 7 + 6 = 2519

314 × 8 + 7 = 2519

279 × 9 + 8 = 2519

251 × 10 + 9 = 2519

vii) 12,345,679 × 9 = 111,111,111

12,345,679 × 18 = 222,222,222

12,345,679 × 27 = 333,333,333

12,345,679 × 36 = 444,444,444

12,345,679× 45 = 555,555,555

12,345,679 × 54 = 666,666,666

12,345,679 × 63 = 777,777,777

12,345,679 × 72 = 888,888,888

12,345,679 × 81 = 999,999,999

viii) 987,654,321 × 9 = 8,888,888,889

987,654,321 × 18 = 17,777,777,778

987,654,321 × 27 = 26,666,666,667

987,654,321 × 36 = 35,555,555,556

987,654,321 × 45 = 44,444,444,445

987,654,321 × 54 = 53,333,333,334

987,654,321 × 63 = 62,222,222,223

987,654,321 × 72 = 71,111,111,112

987,654,321 × 81 = 80,000,000,001

Exercise:

Solve the following by studying the number pattern above:

1) $123,456,789 \times 8 + 9 =$ 2) $123,456,789 \times 9 + 10 =$

3) $98,765,432 \times 9 + 0 =$ 4) $9 \times 9 + 9 + 9 =$

b) Odd Numbers and Squares

When consecutive odd numbers are added together a square pattern emerges.

$1 + 3 = 4 = 2^2$

$1 + 3 + 5 = 9 = 3^2$

$1 + 3 + 5 + 7 = 16 = 4^2$

$1 + 3 + 5 + 7 + 9 = 25 = 5^2$

Exercise:

Write the following sum as power of some number:

1. $1 + 3 + 5 + 7 + 9 + 11 + 13 + 15 =$
2. $1 + 3 + 5 + 7 + 9 + 11 + 13 + 15 + 17 + 19 =$

c) Calculator Addition Patterns

In most of the calculators the numbers 1 to 9 are arranged in the following manner:

7	8	9
4	5	6
1	2	3

Columnwise

741 + 369 = 1110

Rowwise

789+321= 1110

Diagonally

951+159=1110

357+753=1110

Zigzag

478+632=1110

421+689=1110

958+152=1110

d) Pascal's Triangle

Pascal's triangle is a triangular arrangement of numbers in which each number in any row is the sum of the two numbers immediately above it. It is named after Blaise Pascal the French mathematician, although other mathematicians studied it centuries before him in India.

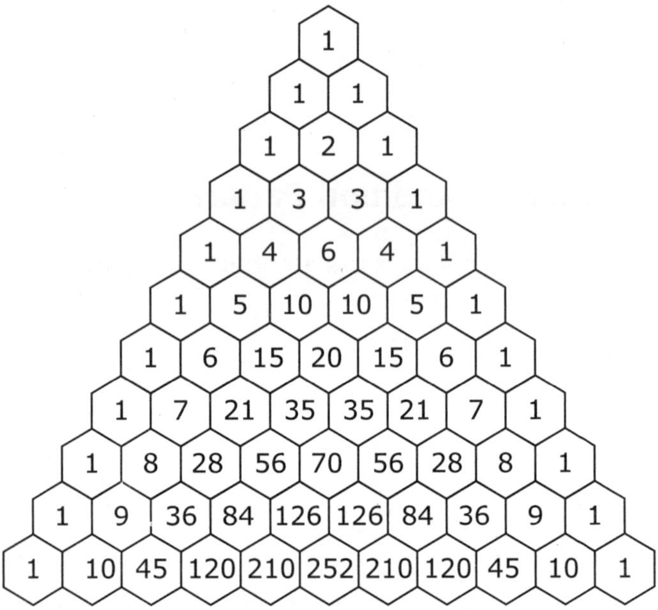

Fig.6.1

Fascinating Patterns of Pascal's Triangle:

i) Diagonals

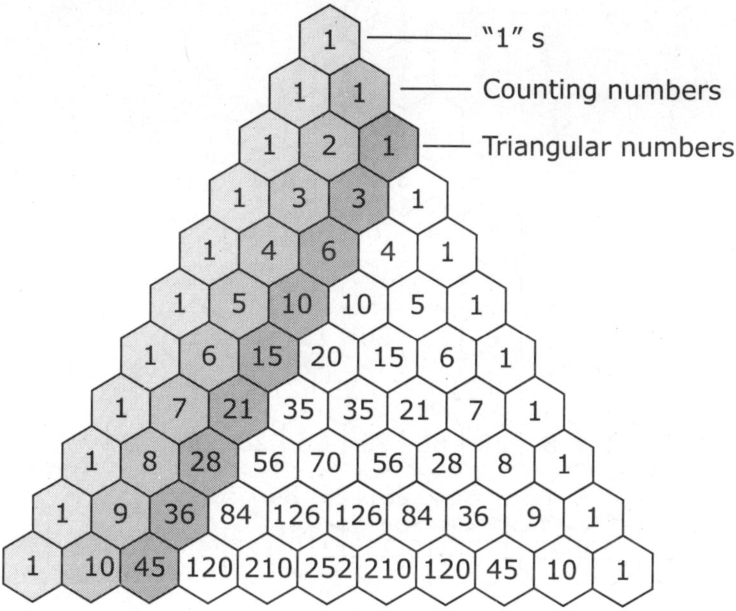

Fig.6.2(a)

A number that can make a triangular dot pattern is called a triangular number. For example,

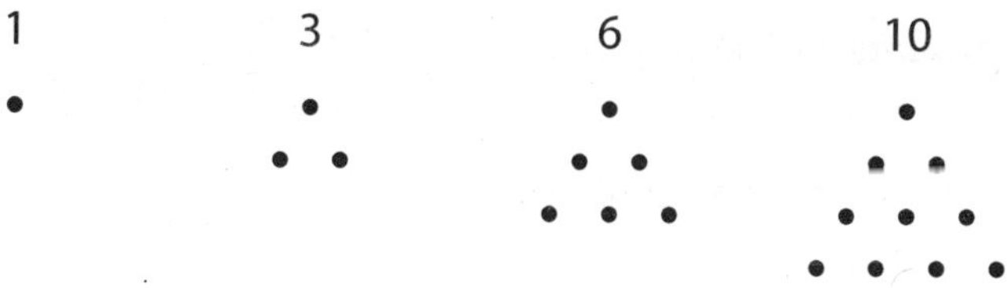

Fig.6.2(b)

ii) Floral pattern

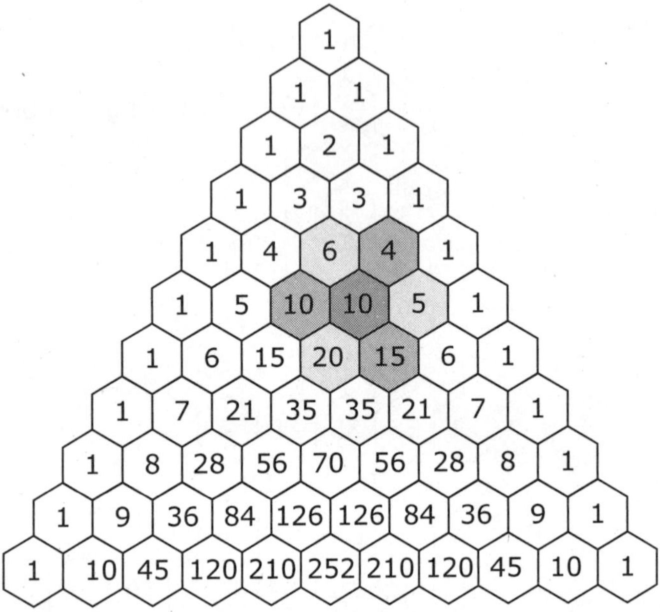

Fig.6.3

Starting with the petal above and to the left of the coloured centre, alternating petals are light coloured and numbered 6, 5 and 20.

The three remaining petals around the coloured centre are dark coloured and numbered 4, 15 and 10.

The product of the numbers in the light petals is

$$6 \times 5 \times 20 = 600.$$

The product of the numbers in the dark petals is

$$4 \times 15 \times 10 = 600.$$

iii) Algebra

Pascal's Triangle gives us the coefficients in binomial expansion:

Power	Polynomial Expansion	Pascal's Triangle
2	$(x+1)^2 = 1x^2 + 2x + 1$	1,2,1
3	$(x+1)^3=1x^3+3x^2+3x+1$	1,3,3,1
4	$(x+1)^4=1x^4+4x^3+6x^2+4x+1$	1,4,6,4,1

Table 6.1

Exercise:

1. Find the sum of numbers in each row of the Pascal's triangle. Do you find any pattern?
2. Can you find the Fibonacci sequence in Pascal's triangle?
3. Does the Pascal's floral pattern work if you start with any cell on Pascal's triangle that is surrounded by 6 other cells?
4. Is there any relation between the number of tosses of a coin and the possible outcomes in Pascal's triangle?
5. Can you find some pattern of 11 in Pascal's triangle? (Hint: Square, cube etc)

+++

7

Amazing facts

a) Circles Around a Circle

You can draw maximum *six* circles (having the same radius) around a circle touching each other. Draw and see for yourself (You may also do it by taking seven coins of the same type).

b) Factors of 1,274,953,680

The number 1,274,953,680 has all the digits from 0 to 9. It is divisible by all the numbers from 1 to 16.

c) Amazing Prime Numbers

Some amazing prime numbers are given below.

31

331

3,331

33,331

However, 333,333,331 is not a prime number. The factors are 17 and 19,607,843.

Exercise:

Write the next two prime numbers in the above sequence.

d) Facts about the Number Seven

1. There are seven wonders of the world.
2. There are seven continents in the world.
3. There are seven colours in the rainbow.
4. There are seven days in a week.
5. There are seven basic musical notes.
6. Most of the mammals' necks have seven bones.
7. The number of rows in the periodic table is seven.
8. The neutral pH value that lies between acidity and alkalinity is seven.

Exercise:

Go to the library or search the internet to find the amazing facts of 23.

e) Strange Multiplication

111,111,111 x 111,111,111 = 12,345,678,987,654,321

12,345,679 x 999,999,999 = 12,345,678,987,654,321

f) Sum of Cubes

The following numbers are equal to the sum of cubes of their digits.

$153 = 1^3 + 5^3 + 3^3$

$370 = 3^3 + 7^3 + 0^3$

Exercise:

Solve the following:

1. $3^3 + 7^3 + 1^3 =$
2. $4^3 + 0^3 + 7^3 =$

g) Division Friends

1 ÷ 37 = 0·027027027... and 1 ÷ 27 = 0·037037037...

h) Reversed Square

$13^2 = 169$

If you find the square of the reversed number the answer is also reversed.

$31^2 = 961$

Exercise:

Find the square of 12 and 21.

i) Multiplication by 9.

1,089 × 9 = 9,801 (The multiplicand is reversed)

10,989 × 9 = 98,901

109,989 × 9 = 989,901

Exercise:

Without actual multiplication find the product of 1,099,989 by 9.

j) Strange Fraction

$$\frac{1}{1089} = 0.00091827364554637281$$

(The numbers in the multiplication table of 9 are 9,18,27,36...)

k) Big Number

The 36,085,288,503,684,007,860 has twenty digits and is divisibly by 20.

The first digit 3 is divisible by 1.

The first two digits 36 is divisible by 2.

The first three digits 360 is divisible by 3.

The first four digits 3608 is divisible by 4.

The first n digits are divisible by n.

Exercise:

See the above pattern and tell by which number are the following divisible:

1. 3,608,528
2. 3,608,528,850,368
3. 36,085,288,503,684,007

l) Mysterious 24

If you square any prime number larger than 3 and subtract 1, the answer will always be divisible by 24.

Example: $19^2 = 361$, $361 - 1 = 360$.

$360 \div 24 = 15$.

Exercise:

Find out the reason why the above division always works.

m) Return to the Original Number

1. Write any number between 0 and 30.
2. Multiply the digit at the ones place by 4 and add the digit at the tens place to the product.
3. Do the same operation with the new number. Keep repeating.
4. Write the numbers in a circle. You will return to the original number within few steps.

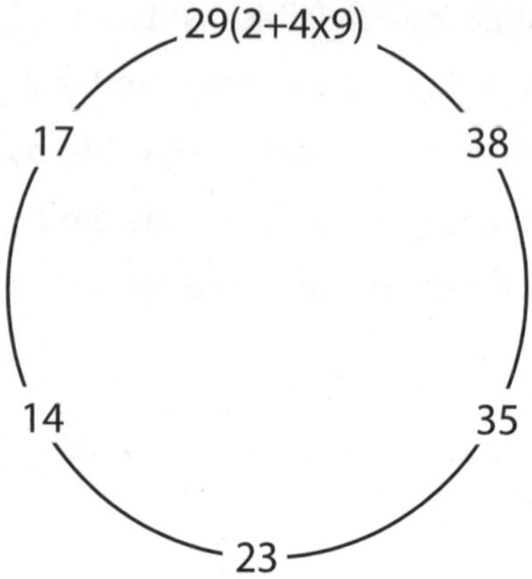

Fig.7.1.

n) The "illions" of Mathematics

Number Name	No. of Zeros followed by 1.
Million	6
Billion	9
Trillion	12
Quadrillion	15
Quintillion	18
Sextillion	21
Septillion	24
Octillion	27
Nonillion	30

Table 7.1

In mathematics, *Googol* is a cardinal number represented as 1 followed by one hundred zeros and *Googolplex* is a cardinal number represented as 1 followed by a googol of zeros (ten raised to the power of a googol).

o) Numerical Palindrome

1. Write any whole number.
2. Reverse its digits and add to the previous number.
3. To this sum add the number formed by reversing the sum's digits.
4. Continue this process. Ultimately you will reach a palindromic number (it reads the same forward and backward).

Example:

1. Let's start with 3659.
2. Reversed number is 9563.
3. Sum of the two numbers is 3659 + 9563 = 13,222.
4. Continuing the process we get 13,222 + 22,231 = 35,453 (palindromic number)

Exercise:

Which is the smallest number that is not known to reach its palindrome when repeatedly added to its reverse?

p) Repeated Numbers

1. Write any four digit number. Repeat the first digit at the end of the number so that it becomes a five digit number.
2. Subtract (smaller from bigger) two consecutive digits (first and second, second and third etc) and write below each pair. Again repeat the first digit of the answer at the end.
3. Continue this process.
4. You will ultimately reach a five digit number with repeated digits.

Example:

Let us take the four digit number as 7859.

7 8 5 9 7 (The first digit 7 is repeated in the end)
  1 3 4 2 1 (Difference of consecutive digits:8-7=1, 8-5=3, 9-5=4, 9-7=2, First digit 1 is repeated)

  2 1 2 1 2
   1 1 1 1 1

q) History of Numbers

The numbers we use (1, 2, 3, etc.) today are known as "Arabic" numbers. Although the Arabs popularized these numbers but they were originally used by the Phoenician merchants to keep track of their trade.

Have you ever thought how the numbers got their names? The Phoenicians used angles to name the numbers.

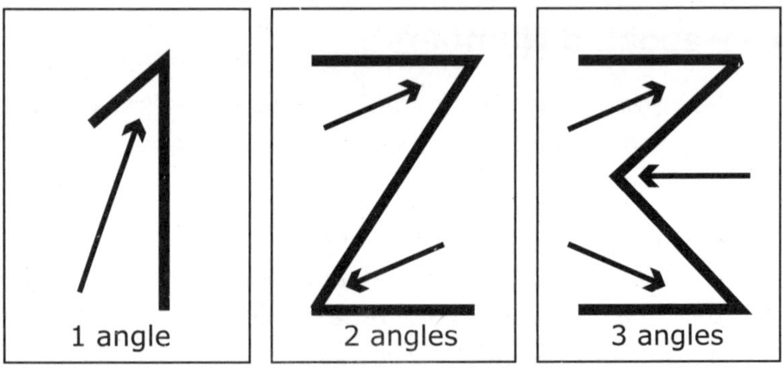

(a) (b) (c)

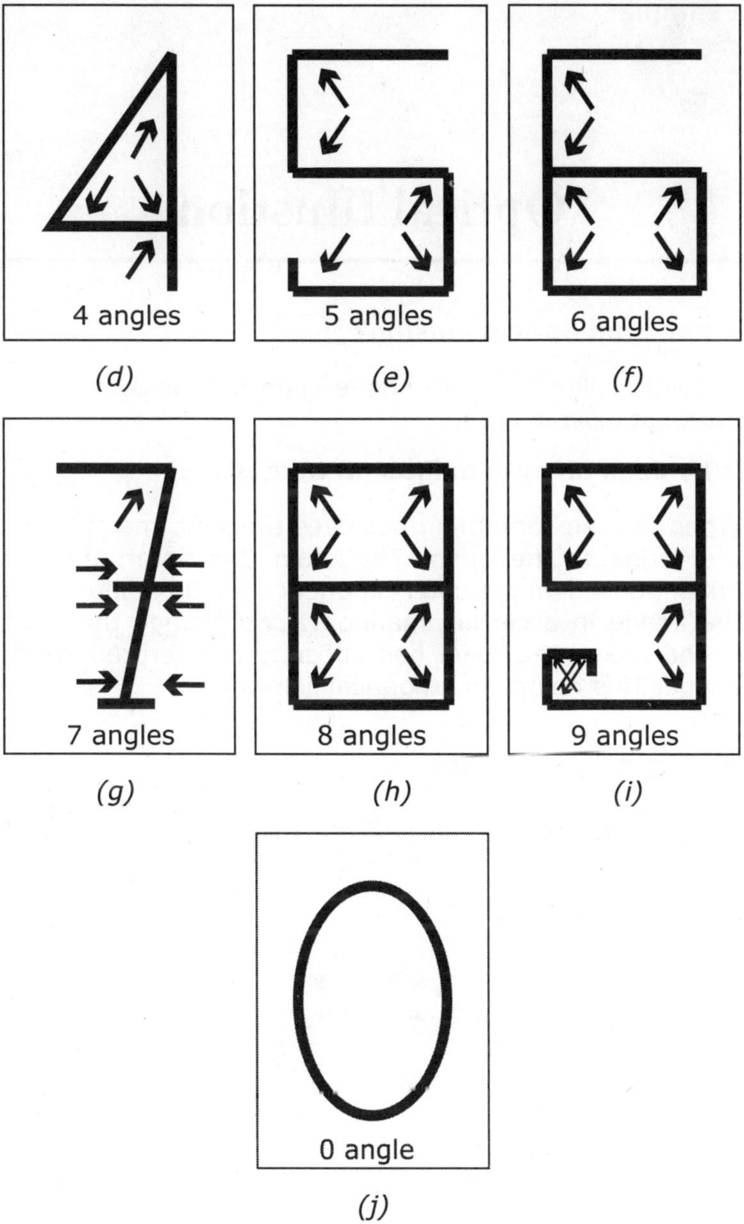

Fig.7.2. Angles in numbers

+++

8

Optical Illusion

What is an optical illusion?

An optical illusion is an image seen by our eyes which does not exist in reality.

Why does an optical illusion take place?

When we see something our eyes transmit images from the retina to the brain. The brain then interprets the image. Based on previous experience the brain interprets the image in a certain manner. If we change the rules of the image the brain has difficulty in interpreting the image. This results in an optical illusion.

a) Rotating Wheel

Stare at the centre (dot) of the following shape for 30 seconds. Move your head in(towards the shape) and out(away from the shape). What do you see?

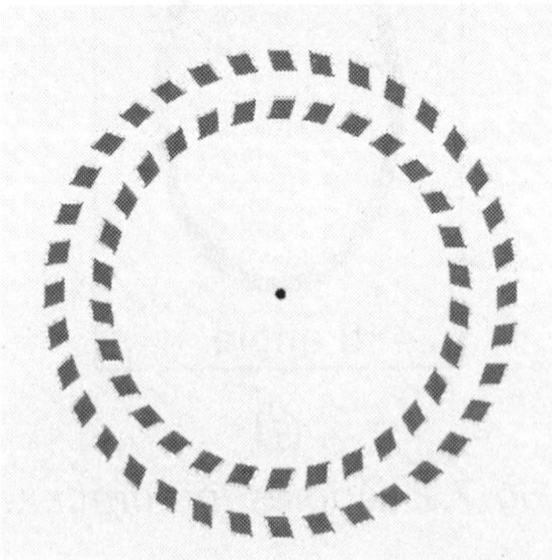

Fig. 8.1

b) Zigzag Wall

Are the rows of the black and white squares parallel or intersecting?

Fig. 8.2

c) Poggendorff illusion

The Poggendorff illusion was discovered by J.C Poggendorf, a physicist in the works of F. Zöllner an astronomer. Which line is a continuation of the black line?

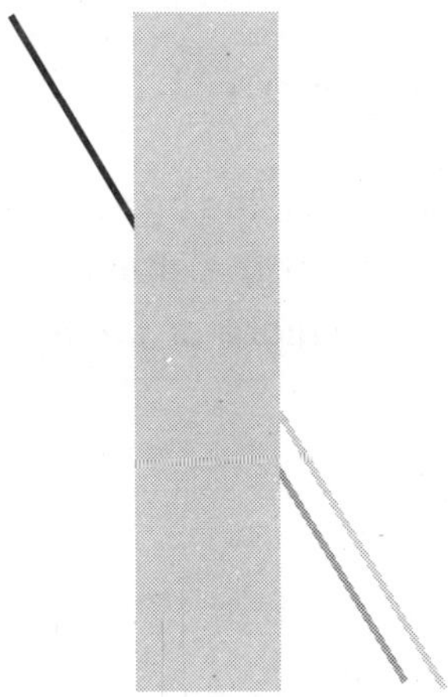

Fig. 8.3

d) Penrose Stairs

The Penrose stairs was created by Lionel Penrose, a British psychiatrist and mathematician and his son Roger Penrose, a British mathematical physicist. Are the stairs ascending (going up) or descending (going down)?

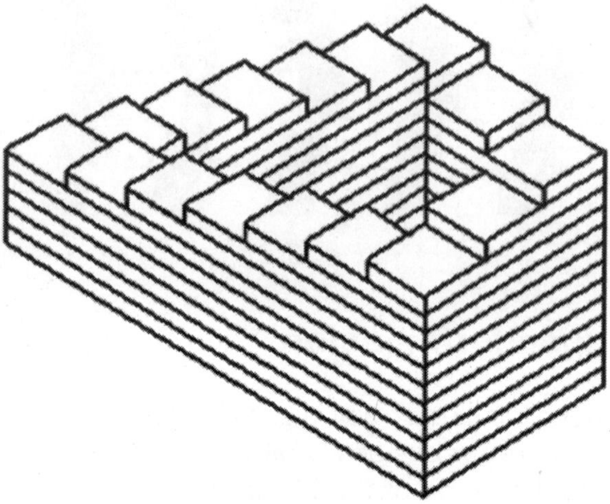

Fig. 8.4

e) Ehrenstein Illusion

This optical illusion was studied by the German psychologist Walter Ehrenstein. Look at the shape inside the concentric circles. Are the sides straight or curved?

Fig. 8.5

Exercise:

In each of the following parts find the shape given in the left in the figure given on the right.

i. 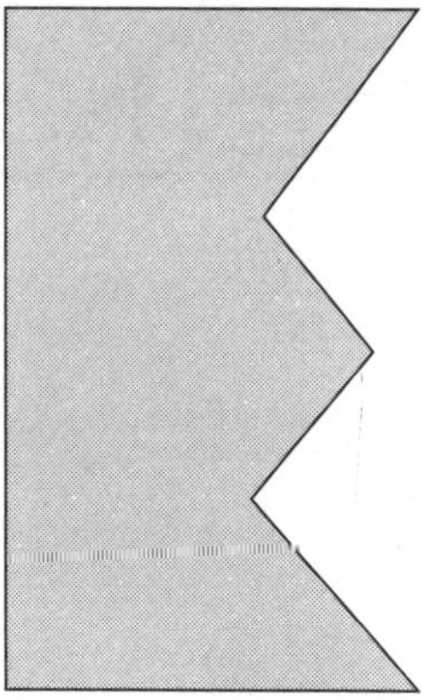

Fig.8.6 (a)

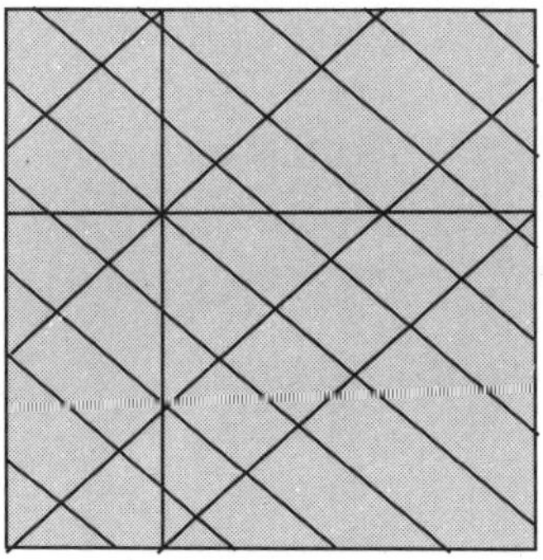

Fig.8.6(b)

ii.

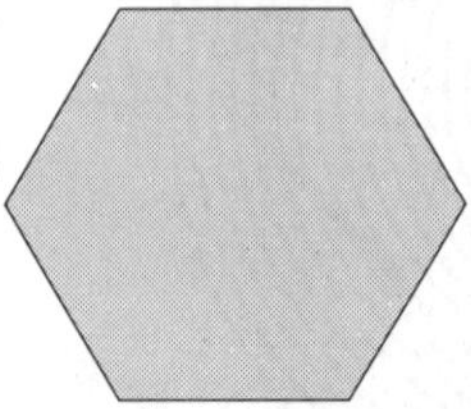

Fig.8.7 (a)

Fig. 8.7 (b)

iii.

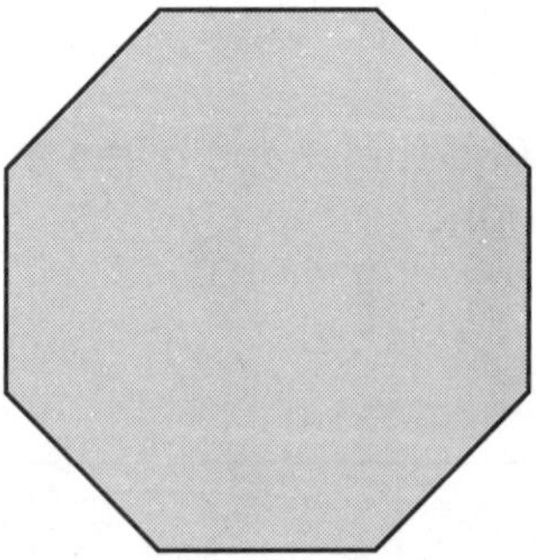

Fig.8.8(a)

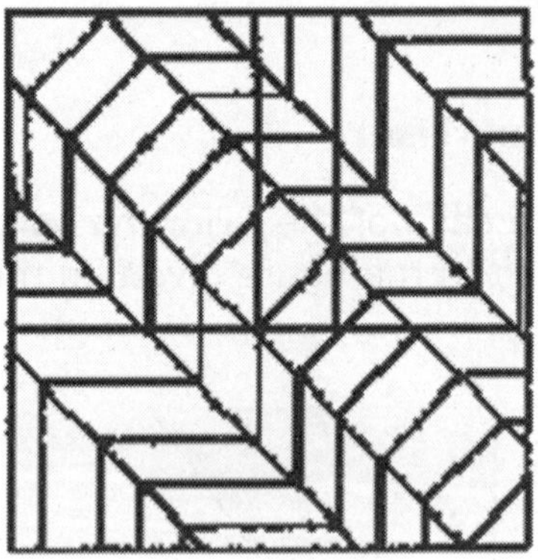

Fig.8.8(b)

+++

9

Mathematics in everyday life

Mathematics is a part of our daily life. We use mathematics in most of our activities everyday. For example time, date, money, address, phone numbers, banking, buying and selling goods etc are all parts of mathematics. Apart from this we have mathematics in every sphere of life like nature, art, architecture, medicine, textile, sports, music, cooking etc.

a) Nature

We find various concepts of mathematics (shape, size, symmetry, tessellation, Fibonacci sequence, Golden ratio etc) in nature. The seeds, leaves, fruits, flowers appear in different shapes and sizes.

(a) Bilateral symmetry

(b) Radial symmetry

(c) Tessellation

Most of the flowers have Fibonacci number of petals. The scales of pineapple and bracts of pinecone also follow the Fibonacci concept.

The *nautilus shell* follows the concept of Golden spiral.

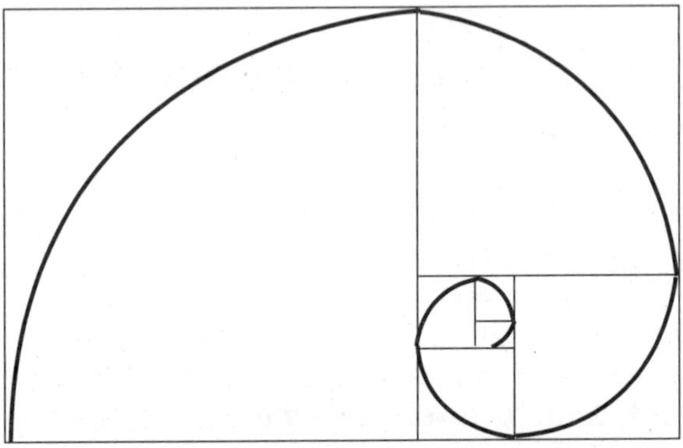

(d) Nautilus Shell (use of Golden spiral)

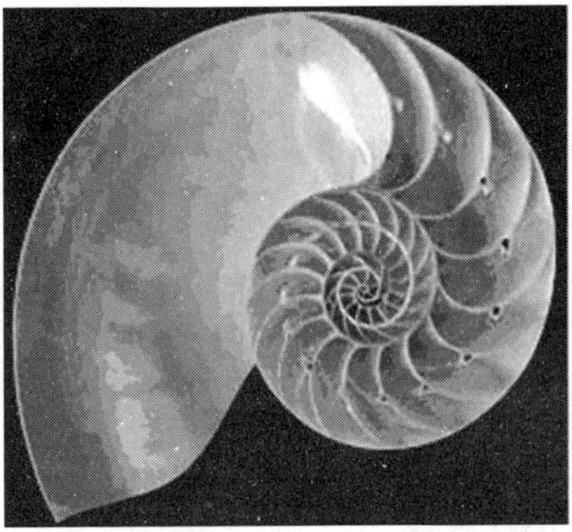

(e) Nautilus Shell

Fig. 9.1

It seems even bees understand Geometry! The beehives are always hexagon in shape because a hexagon encloses the largest possible area and takes the smallest perimeter to enclose that area. Hence the bees can use less building material and less energy to make their homes. Moreover, hexagon can be easily tessellated.

b) Architecture

Mathematics forms the base of architecture. In both ancient as well as modern architecture various mathematical concepts like measurements, area, volume, angle, coordinate geometry etc are used. Various ancient architectural wonders like the great wall of China, the Pyramids of Egypt, the Parthenon, the Colosseum and the Taj Mahal etc were all built using the fundamental principles of mathematics. The ancient architects also used the concept of Fibonacci, Golden mean, Pythagoras theorem, Pi etc effectively.

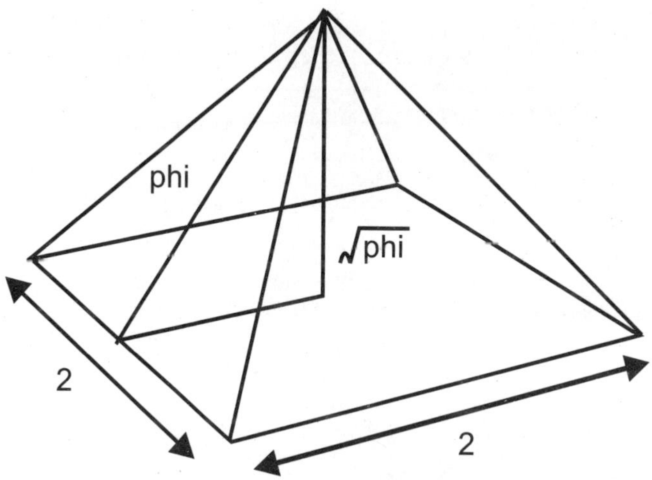

(a) Pyramids of Egypt (use of Golden Ratio)

(b) Pyramids of Egypt

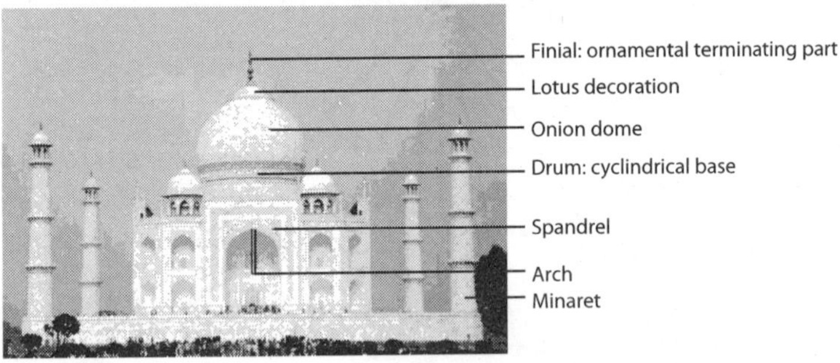

(c) Taj Mahal (Agra)

Fig.9.2

c) Music and Dance

One of the earliest known applications of mathematics in music is attributed to Pythagoras. One day Pythagoras was walking by the side of few blacksmiths when he heard the sound of the hammers on the anvils. After a short while, he noted that one of the hammers was sounding in harmony. He made a thorough examination of the hammers and discovered that when their masses were simple ratios i.e. 2:1 or 4:1, then the respective notes produced were in harmony. This illustrates the first application of mathematics in music.

Later the work of Pythagoras was improved by Galilei (the father of the famous Galileo Galilei) who concluded that the best frequencies were in the following proportions :

Sa	re	ga	ma	pa	dha	ni	sa
do	re	mi	fa	so	la	ti	do
1	9/8	5/4	4/3	3/2	5/3	15/8	2

which may be represented as whole number proportions as

24 27 30 32 36 40 45 48

Mathematics also plays an important role in the construction of musical instruments.

Not only music but even dance involves mathematics. Every rhythm(every dance step) is a number. It follows a specific pattern and symmetry.

d) Art

Art does not always require the four fundamental operations of mathematics but it uses the concept of various geometric shapes, symmetry, translation, tessellation etc.

In the following painting by M C Esher we can see the concept of tessellation. (Notice the black and white horses facing each other)

(a) Horsemen by M C Esher

The famous artist Leonardo da vinci used the concept of Golden rectangle in his famous painting Mona Lisa.

(b) Mona Lisa by Leonardo da Vinci

Fig.9.3

e) Sports

Every sport uses mathematics in some way or the other. The number of players, recording of score, determining the average or percentage score, graphical representation of score of various teams, dimension of the sports ground/court/pitch/ring/pool etc. everything involves mathematics. The concept of angle is also very important in sports.

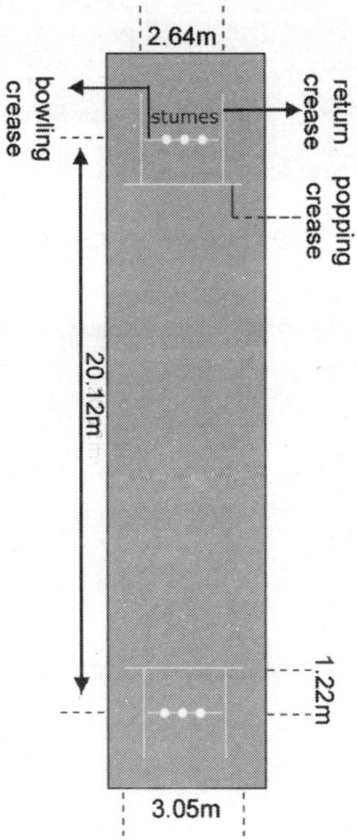

Fig.9.4 Dimensions of turf pitch of cricket

f) Textile

We can find exotic geometric designs in both tapestry (bedcover, cushion cover, curtain, quilt etc) and textiles(dress material, saree etc). Point, parallel lines, intersecting lines, angles, various shapes etc are the major concepts used in this field.

(a) Cushion cover

(b) Saree

Fig.9.5

g) Decoration

Geometry is extensively used in handicrafts, jewellery and rangoli (the art of drawing motifs and images on the floor of one's home using rice powder, flower and different coloured powders).

(a) Wall hanging

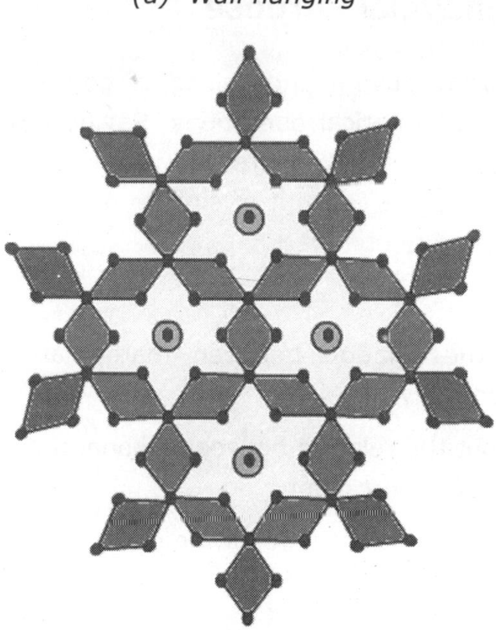

(b) Rangoli

Fig.9.6

h) Cooking

Without realizing we use various mathematical concepts in our everyday cooking. Measurement, ratio, proportion, fraction etc are some of the concepts used while cooking food.

i) Weather Prediction

Global warming and its effect on the weather is a major concern in today's world. The complex interaction between the polar ice caps, oceans and the atmosphere is studied using statistics, probability, calculus and differential equations.

j) Commercial Purpose

Our everyday buying and selling of goods involves all the four mathematical operations. Saving money in the form of bonds, stock, shares or just keeping it in the bank involves the concept of simple interest, compound interest, annuity etc.

Exercise:

1. Find the relation between mathematics and the human body.
2. Find out the relation between Fibonacci numbers and the *sneezewort* plant.
3. Draw and colour the following rangoli patterns:

(a)

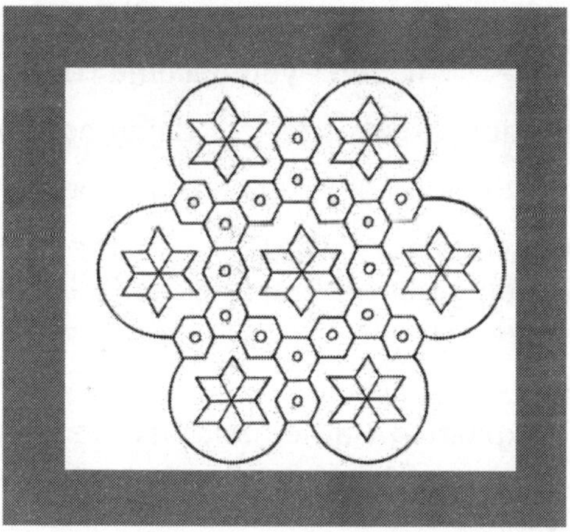

(b)

Fig.9.7

+++

10

Mathematical Jokes

a. What did zero say to eight?

You have a nice belt.

b. Why is six scared of seven?

As seven eight(ate) nine.

c. What did "*i*" and "*pi*" say to each other?

"*i*" to "*pi*" – "You better be rational."

"*pi*" to "*i*" - "Why don't you become real?"

d. Mathematical excuses for not doing homework:

i. I was thinking of imaginary numbers.

ii. I was tending towards my mathematics book. I could only get arbitrarily close to it but I couldn't actually reach it.

e. Mathematics teacher: What is the full form of Maths?(expecting Mathematics as the answer)

Student: Mentally Affected Teachers Harassing Students.

f. Ritika: "Shreya, please lend me your algebra homework."

Shreya: "How can I ? Sir will scold us both for cheating."

Ritika: "Don't worry, I will change all the variables."

After the class Shreya asked Ritika "Did you change the variables?"

Ritika: "Yes Of course! I changed the entire *x* to *a*, *y* to *b*, *z* to *c* and *log* to *timber*.

g. Geometry teacher: Who can draw the shortest distance between points M and N drawn on the blackboard?

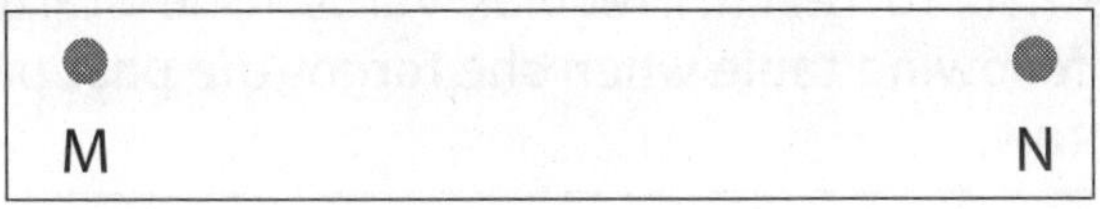

Fig.10.1(a)

Student: It is very easy madam! He draws the following:

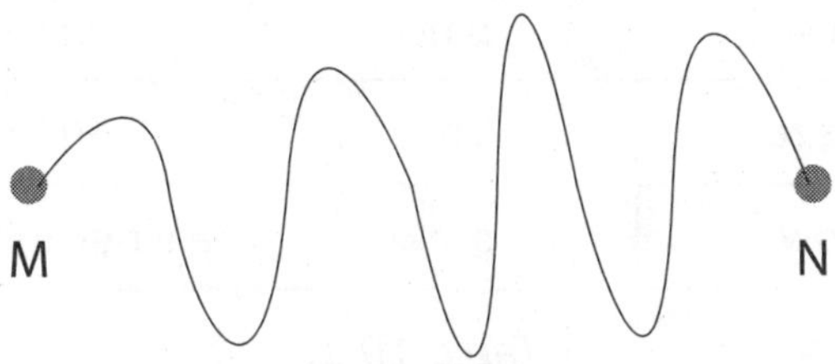

Fig.10.1(b)

Shocked teacher: Who taught you this?

Student: My father, he is an auto rickshaw driver.

h. In a calculus class the teacher asks the students to find the following limit.

$$\lim_{y \to 2} \frac{1}{y-2}$$

One of the students write

$$\lim_{y \to 2} \frac{1}{y-2} = \text{∾}$$

The teacher is surprised and teils the student to explain the answer.

He says "Sir, I solved it according to the example given by you in the previous class."

Teacher's example in the previous class:

$$\lim_{y \to 8} \frac{1}{y-8} = \infty$$

i. One day the English teacher was absent and the Mathematics teacher was asked to take her class. She had to teach irregular verbs. She started with the following table when she forgot the past participle of grow.

Infinitive	Simple Past	Past Participle
begin	began	begun
blow	blew	blown
freeze	froze	frozen
grow	grew	Teacher forgot this

Table 10.1.

Then she said let the unknown quantity be *x*. We will derive it by using proportion.

$$\frac{blew}{grew} = \frac{blown}{x}$$

$$Or, x = \frac{grew.blown}{blew}$$

Cancelling the common terms we get,

$$x = \frac{gr\cancel{ew}.\cancel{bl}own}{\cancel{blew}}$$

$x =$ grown

Hence the past participle of grow is grown.

Funny Proofs:

a. *Given:* 1 Re = 100 p where Re stands for rupee and p stands for paisa.

To Prove: 1 Re = 1p

Proof:

	1 Re = 100p	(Given)
Or	1 Re = $(10p)^2$	[100p = $(10p)^2$]
Or	1 Re = $(0.1Re)^2$	(10p = 0.1 Re)
Or	1 Re = 0.01Re	[$(0.1)^2$ = 0.01]
Or	1 Re = 1p	(0.01Re = 1p)

b. *Given:* If you study you won't fail otherwise you will fail.

To Prove: Study = Fail

Proof:

Study = Not fail (Given) (1)

Not study = Fail (Given) (2)

Adding equation (1) and (2) we get:

Study + Not Study = Not fail + fail

Factorizing from both sides we get:

Study(1 + Not) = Fail(1 + Not)

Dividing both sides by (1 + Not) we get:

Study = Fail

c. *Given:* One of the doors of a room is half closed.

To Prove: Open door = Closed door

Proof:

$$\frac{1}{2}\textit{Open door} = \frac{1}{2}\textit{Close door}\ (\textit{Given})$$

Multiplying both sides by 2 we get:

Open door = Closed door

d. *Given:* Time is money. Knowledge is power.

To Prove: The lesser knowledge one has the greater money one can make.

Proof:

$$Power = \frac{Work\ done}{Time\ taken} (Scientific\ fact)$$

Since Power = Knowledge and Time = Money we get,

$$Knowledge = \frac{Work\ done}{Money}$$

$$or,\ Money = \frac{Work\ done}{Knowledge}$$

Since money is inversely proportional to knowledge hence if knowledge decreases money increases irrespective of the work done.

e. *Given:* log (1) = 0

To Prove: log (-1) = 0

Proof:

$\log(-1)^2 = 2\log(-1)$ (since $\log m^n = n \log m$) (1)

Also $\log(-1)^2 = \log(1)$ [since $(-1)^2 = 1$] (2)

Subtracting equation (2) from equation (1) we get,

$2\log(-1) - \log(1) = 0$

Or, $2\log(-1) = \log(1)$

Or, $2\log(-1) = 0$ [since log (1) = 0]

Dividing both sides by 2 we get,

$\log(-1) = 0$

11

Secret of Success in Mathematics

To find out the secret of being successful in mathematics let us assign numbers to alphabets according to the following table:

Alphabet	Number	Alphabet	Number
A	1	N	14
B	2	O	15
C	3	P	16
D	4	Q	17
E	5	R	18
F	6	S	19
G	7	T	20
H	8	U	21
I	9	V	22
J	10	W	23
K	11	X	24
L	12	Y	25
M	13	Z	26

Table 11.1

Now let us calculate the percentage of some words.

K – N – O – W – L – E – D – G – E

11 + 14 + 15 + 23 + 12 + 5 + 4 + 7 + 5 = 96%

H – A – R –D – W – O – R – K

8 + 1 + 18 + 4 + 23 + 15 + 18 + 11 = 98%

A – T – T –I – T – U – D – E

1 + 20 + 20 + 9 + 20 + 21 + 4 + 5 = 100%

Thus, it is proved with mathematical certainty that after acquiring knowledge if we do hardwork and have the right (positive) attitude we will be successful in mathematics.

Answers

Mathematical Activities (Chapter 3)

a) Permutation

You will get 6 * 6 * 6 = 216 permutations.

b) Paper Folding

Number of times folded (n)	Number of sections (2^n)	Number of creases (2^n-1)
1	$2 = 2^1$	1
2	$4 = 2^2$	3
3	$8 = 2^3$	7

Table 12.1

As per the pattern if we could fold the paper nine times we would get $2^9 = 512$ sections and 511 creases.

c) Regular Polyhedra

The following table shows the number of faces, edges and vertices of the platonic solids.

Shapes	Number of Faces	Number of Edges	Number of Vertices
Tetrahedron	4	6	4
Cube	6	12	8
Octahedron	8	12	6
Dodecahedron	12	30	20
Icosahedron	20	30	12

Table 12.2

e) Möbius Strip

When we cut the loop along the length we get a loop bigger than the original one. Möbius strip is part of a branch of mathematics called topology.

f) Pythagoras Theorem

- Total small squares in a square of side 3cm + Total small squares in a square of side 4cm = Total small squares in a square of side 5cm
- In a right triangle the square of the hypotenuse is equal to the sum of the square of the other two sides.
- The Pythagoras theorem is valid only in a right angled triangle.

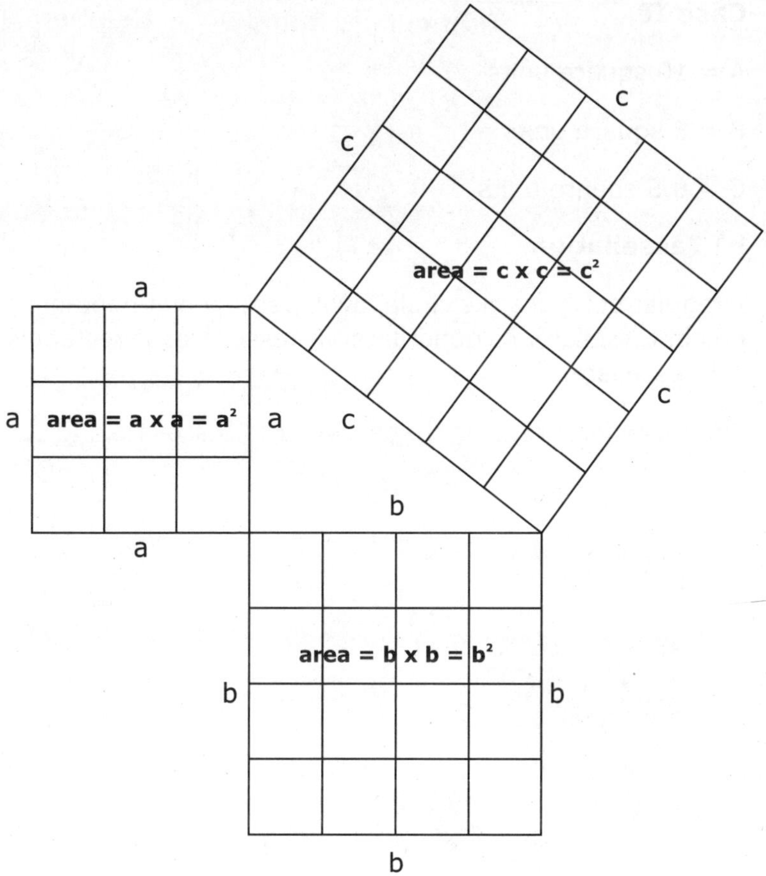

Fig. 12.1

In the above example, 9 + 16 = 25 or $3^2 + 4^2 = 5^2$ or $a^2 + b^2 = c^2$.

g) Geoboard

Case I

Six

Fig. 12.2

Case II

A = 10 square units

B = 8 square units

C = 8.5 square units

h) Tessellation

Irregular polygons like circle, oval, pentagon or heptagons do not tessellate. All quadrilaterals tessellate, so rectangle will tessellate.

The following irregular shapes will also tessellate.

Fig. 12.3

i) String Art

Case I

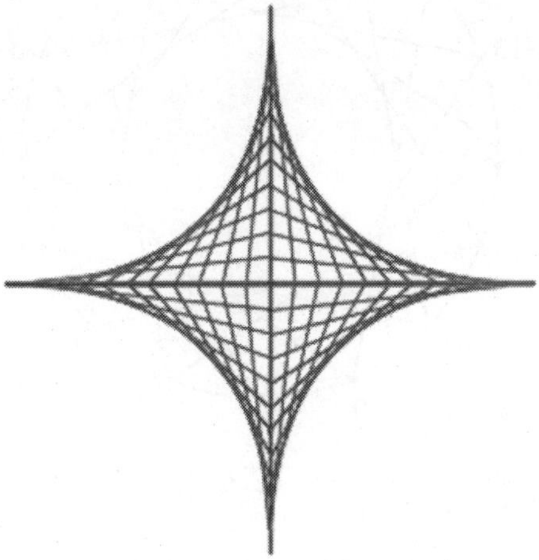

Fig. 12.4(a)

Case II

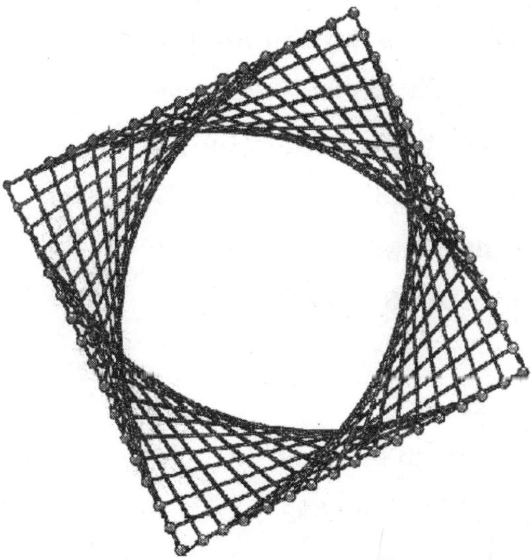

Fig.12.4(b)

Case III

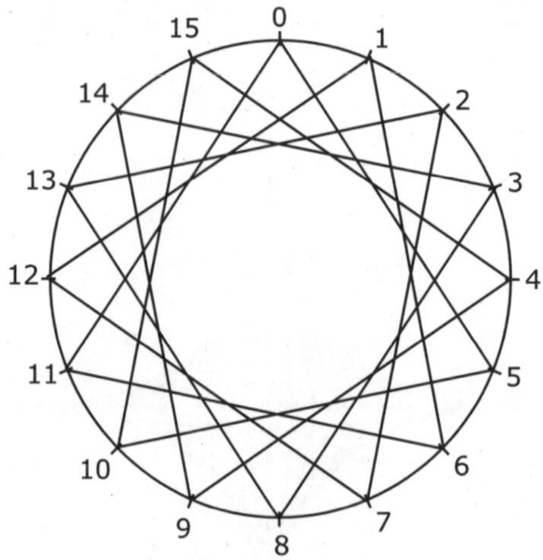

Fig. 12.4(c)

Trick and Quick (Chapter 4)

a to c) Squaring of numbers

1) 3249 2) 6889 3)11881 4)17424 5)8836

g) Multiplying two digit numbers by 11

1)319 2)968 3) 517

h) Multiplying big numbers by 11

1)38016 2) 132440 3) 95579

i) Some Multiplication Rules

1) 522 2)2028 3)660

4)833 5)5092 6)7425

j)Time Constant

1) 16 hours 20 minutes

2) 6 hours 50 minutes

k) Easy Division by 5

1) 877.8 2)2313.4 3)14190.6 4)17799.4

l) Easy Division by 9

1) 0.88... 2)0.0404... 3)0.497497... 4)0.067067...
5) 0.008008...

m) Perfect Number

496, 8128, 2096128

o) Quick division by 9

1. Quotient = 6, Remainder = 3
2. Quotient = 10, Remainder = 6
3. Quotient = 509, Remainder = 6
4. Quotient = 1210, Remainder = 7

p) Vedic Mathematics

Exercise:

A. Use *Nikhilam Navatashcaramam Dashatah* to solve the following:

1. 113 2. 31025 3. 9212
4. 9215 5. 11128 6. 11227

B. Use *Ekadhikena purvena sutra* to solve the following:

1. 5625 2. 7225
3. 0.0344827586206896551724137931
4. 0.025641025641025641

C. Use *Urdhva Tiryagbhyam sutra* to solve the following:

1. 3818 2. 2886 3. 3604 4. 3471

D. Use *Anurupye Sunyamanyat sutra* to solve the following equations:

i. $x = 0, y = 2/7$ ii. $x = 2, y = 0$

E. Use *Sankalana Vyavakalanabhyam* to solve the following equations:

i. $x = 2, y = -1$ ii. $y = 2, z = 3$

Puzzles (Chapter 5)

a. Compass Direction

0

b. Proof of 1 = 2 using algebra.

The division by (a – b) in step 5 is not correct since a – b is zero. Division by zero is not defined.

c. Proof of π (pi) = 3 using algebra

We cannot always take the positive square root. In this case when we take the square root

$$-(x - \pi) = +(x - 3)$$

d. Alphamagic Square

5	22	18
28	15	2
12	8	25

Table 12.3

In the first row of the above table the number of letters in the word five is 4, the number of letters in the word twenty two is 9, the number of letters in the word eighteen is 8 etc.

4 (five)	9 (twenty two)	8 (eighteen)
11 (twenty eight)	7 (fifteen)	3 (two)
6 (twelve)	5 (eight)	10 (twenty five)

Table 12.4

e. T Puzzle

Fig. 12.5

f. Doctor's Medicine

One hour. If Sunil takes the first pill at 1:00, the second at 1:30, and the third at 2:00, the tablets are finished and only one hour has passed.

g. Same Result

5

h. Coloured Wrist Band

The two options for Mr. Blue are white or red wrist band. But the person wearing red wrist band said he was right. Hence the only option for Mr. Blue is white. Mr.White is wearing red wrist band and Mr.Red blue.

i. Cut the Cake

She should make the first two cuts like a + from the top of the cake. Next she should slice the cake horizontally.

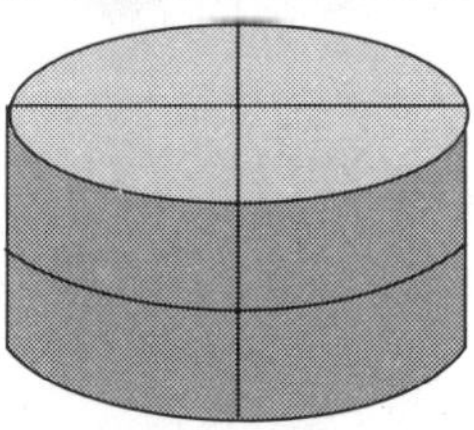

Fig. 12.6

j. Number Puzzle

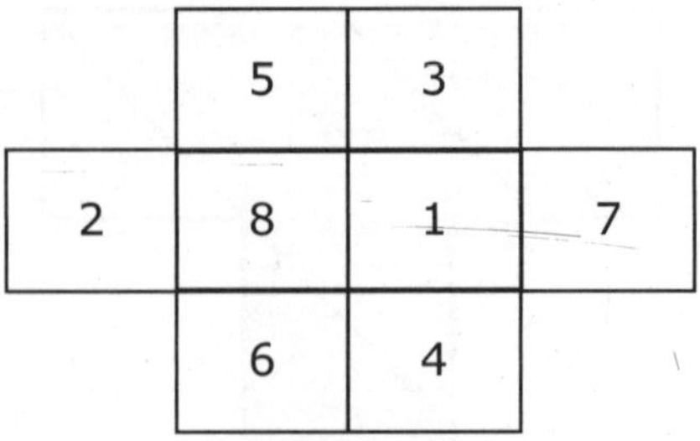

Fig. 12.7

k. Oldest Number Puzzle

One. Only the speaker was going to St.Ives. He met the remaining people on his way.

l. Roman Numbers

I = III – II

V + IV = IX

m. Friar's Puzzle

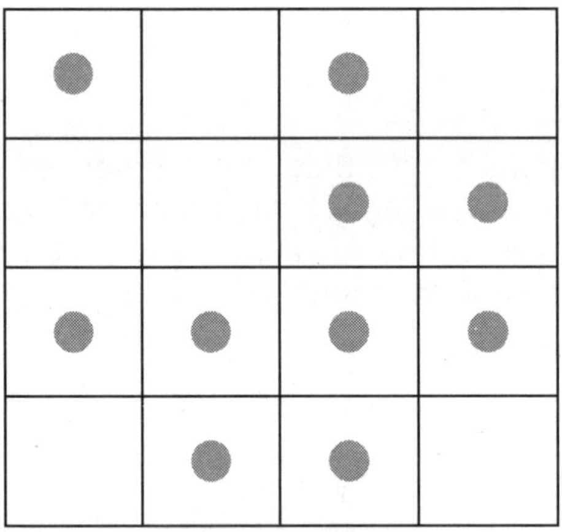

Fig. 12.8(a)

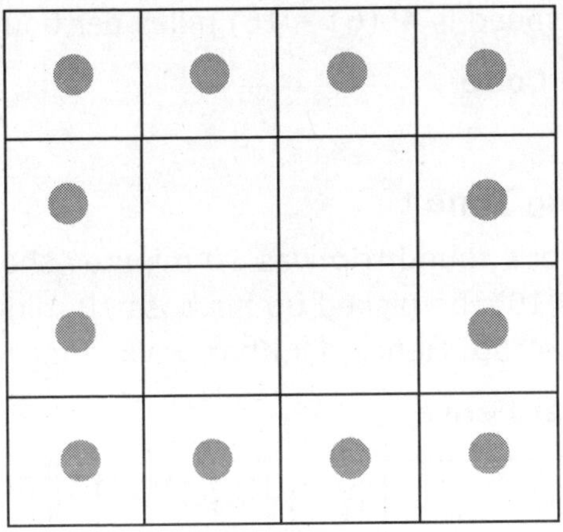

Fig. 12.8(b)

n. Car Speed

45 miles per hour

Let the number on the 1st mile stone = $10x + y$

After an hour, the number on the 2nd mile stone = $10y+x$

An hour later the number on the 3rd mile stone = $100x + 0 + y$

Since he crosses the mile stone after every hour with a uniform speed,

Distance covered in 1st hour = Distance covered in the 2nd hour

$(10y + x) - (10x + y) = (100x + y) - (10y + x)$,

by solving we get,

$y = 6x$ or $x/y = 1/6$

When $x = 1$, $y = 6$

All other values are not possible as we end up with a double digit value for y, as x varies.

Hence the speed is 45(61 – 16) miles per hour.

o. Morse Code

Forty

p. Walking Time

As the driver's round trip was 20 minutes shorter ($x + x = 20$ or $x = 10$), he picked up his boss 10 minutes before 5:00 i.e. at 4:50. Hence, Mr.Khan walked for 50 minutes.

q. Number Fence

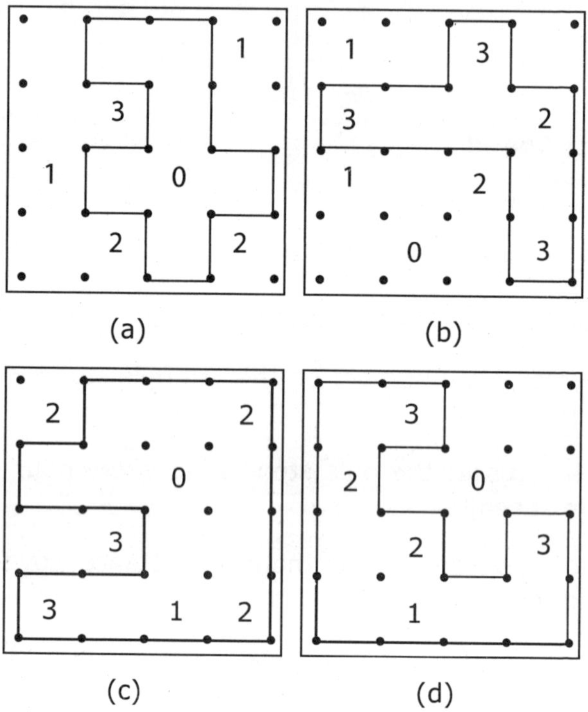

Fig. 12.9

r. Grandmother's Gift

Loose shoe. Key in the numbers 35007 and 3045 one at a time and hold the calculator upside down.

s. Burning Rope

Light the first rope from both the ends and the second rope from one end. When the first rope is finished burning 30 minutes are over and the second has 30 minutes remaining.

Now, light the other end of the second rope so that remaining part of it will burn in 15 minutes. Thus 30+15=45.

t. Age

The two teachers were 25 and 24.

$25=5^2$, the next square is $6^2=36$, i.e. in 11 years.

$24=4!$, the next factorial is $5!=120$, i.e. in 96 years.

u. Tangram

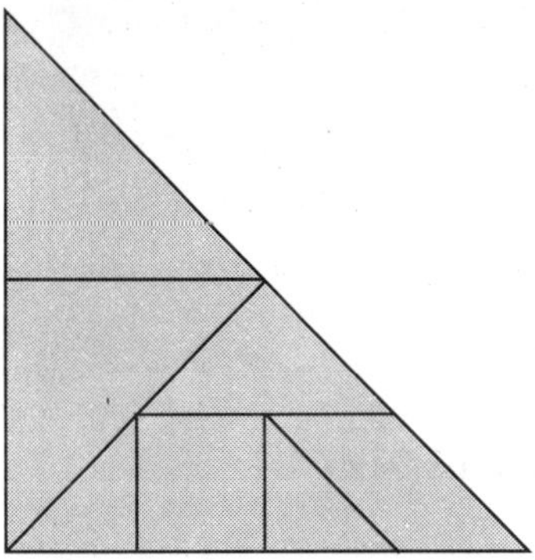

(a)

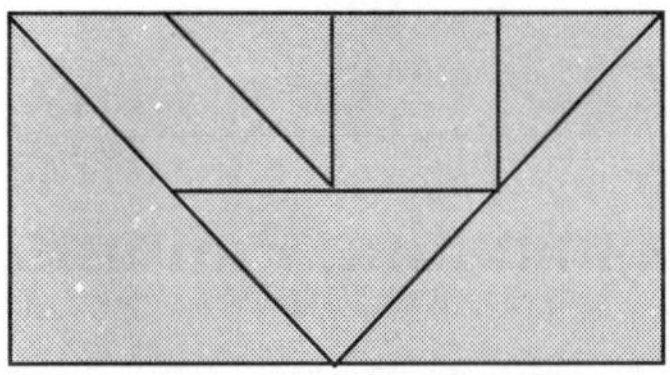

(b)

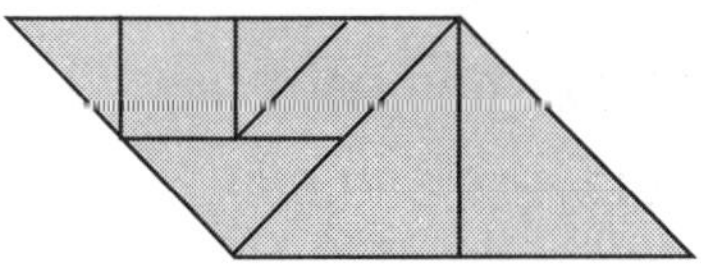

(c)

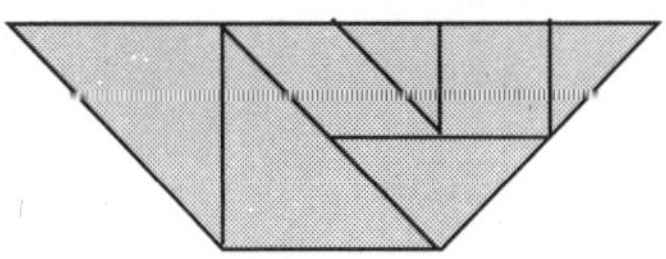

(d)

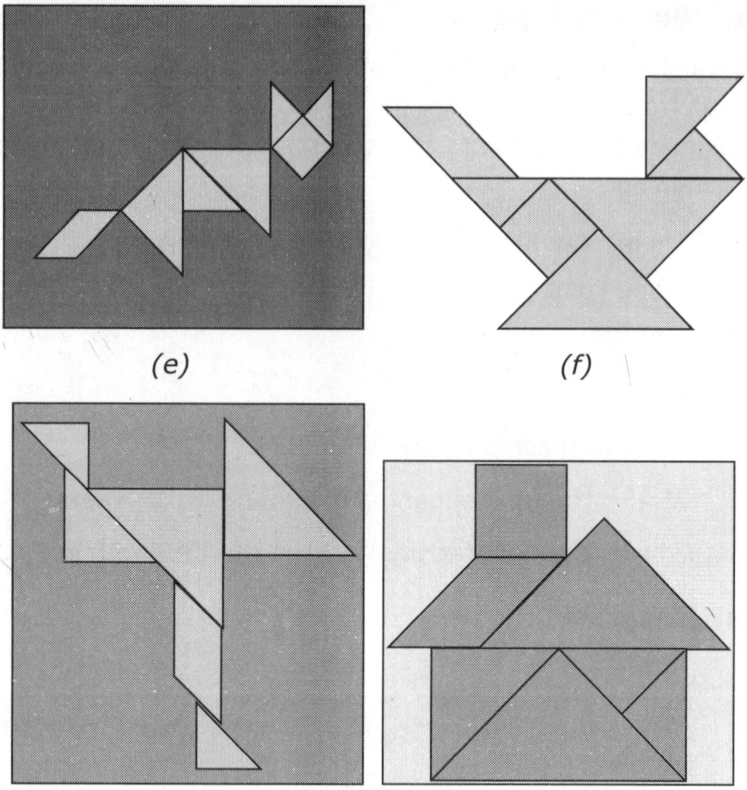

(e) (f)

(g) (h)

Fig. 12.10

Fascinating Patterns (Chapter 6)

a. Number Patterns

Exercise I

1. 123,454,321
2. 12,345,654,321
3. 1,111,088,889
4. 111,110,888,889

Exercise II

1. 123456789 × 8 + 9 = 987,654,321
2. 123456789 × 9 + 10 = 1,111,111,111

3. $98765432 \times 9 + 0 = 888,888,888$

4. $9 \times 9 + 9 + 9 = 99$

$$9 \times 9 = 81$$
$$99 \times 99 = 9801$$
$$999 \times 999 = 998001$$
$$9999 \times 9999 = 99980001$$
$$99999 \times 99999 = 9999800001$$
$$999999 \times 999999 = 999998000001$$
$$9999999 \times 9999999 = 99999980000001$$
$$99999999 \times 99999999 = 9999999800000001$$
$$999999999 \times 999999999 = 999999998000000001$$

b. Odd Numbers and Squares

1. $1 + 3 + 5 + 7 + 9 + 11 + 13 + 15 = 64 = 8^2$
2. $1 + 3 + 5 + 7 + 9 + 11 + 13 + 15 + 17 + 19 = 100 = 10^2$

c. Pascal's Triangle

1. The sum of the numbers in each row of the Pascal's Triangle is as follows :

   $1 = 2^0$

   $2 = 2^1$

   $4 = 2^2$

   $8 = 2^3$

2. The numbers on the diagonal of the triangle add to Fibonacci Sequence: 1, 1, 2, 3, 5, 8, 13, 21, 34, 55, 89, 144, 233, ...

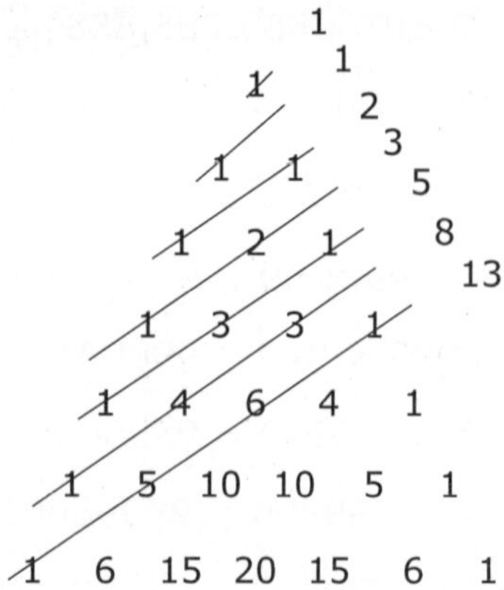

Fig. 12.11

3. Yes the Pascal's floral pattern works if we start with any cell on Pascal's triangle that is surrounded by 6 other cells.

4. The relation between the number of tosses of a coin and the possible outcomes in Pascal's Triangle is as follows:

Number of tosses	**Possible outcome**	**Pascal's Triangle**
1	T H	1,1
2	TT TH,HT HH	1,2,1
3	TTT TTH,THT,HTT HHT, HTH, THH HHH	1, 3, 3, 1
4	TTTT THHH, HHTH, HHHT, HTHH THTH, HHTT, HTHT, HTTH, THHT, TTHH TTTH, THTT, TTHT, HTTT HHHH	1,4,6,4,1

Table 12.5

1. Some pattern of 11 in the Pascal's Triangle is as follows:

Row#	Formula	Result	Actual Row
Row 0	11^0	1	1
Row 1	11^1	11	11
Row 2	11^2	121	121
Row 3	11^3	1331	1331
Row 4	11^4	14641	14641

Table 12.6

Amazing facts (Chapter 7)

c. Amazing Prime Numbers

333,331 and 3,333,331

d. Facts about the Number Twenty Three

1. Charles Darwin's Origin of Species was published in 1859 (1+8+5+9 = 23).
2. Each parent contributes 23 chromosomes to start the human life.
3. The Hiroshima bomb was dropped around 8.15am (8+15= 23).
4. The 11th September, 2001 attack on twin tower (USA) adds up to 23 (11 + 9 + 2 + 0 + 0 + 1 = 23)
5. The Titanic ship sank on 15th April 1912(1+5+4+1+9+1+2=23).
6. The "W" on the keyboard is right between 2 and 3.
7. The tilt of Earth's axis is roughly 23 degrees.
8. The average circumference of a human head in inches is 23.

f. Sum of Cubes

1. $3^3 + 7^3 + 1^3 = 371$ 2) $4^3 + 0^3 + 7^3 = 407$

h. Reversed Square

$12^2 = 144$ and $21^2 = 441$

i. Multiplication by 9

9899901

k. Big Number

1) 7 2) 13 3) 17

l. Mysterious 24

All prime numbers can be written as $(6n+1)$ or $(6n-1)$.

$(6n + 1)^2 = 36n^2 + 12n + 1$

$(6n + 1)^2 - 1 = 36n^2 + 12n$

$(6n + 1)^2 - 1 = 12n(3n + 1)$

Either n or (3n + 1) must be even, hence 12n(3n + 1) is divisible by 24.

$(6n - 1)^2 = 36n^2 - 12n + 1$

$(6n - 1)^2 - 1 = 36n^2 - 12n$

$(6n - 1)^2 - 1 = 12n(3n - 1)$

Either n or $(3n - 1)$ must be even, hence $12n(3n - 1)$ is divisible by 24.

o. Numerical Palindrome

196

Optical Illusion (Chapter 8)

a. Rotating wheel

You will see that the inner wheel rotates.

b. Zigzag wall

The rows of black and white squares are all parallel. Check it with a ruler! The vertical zigzag patterns disrupt our horizontal perception.

c. Poggendorf illusion

It appears that the dark coloured line is a continuation of the black line but actually the light coloured line is the continuation.

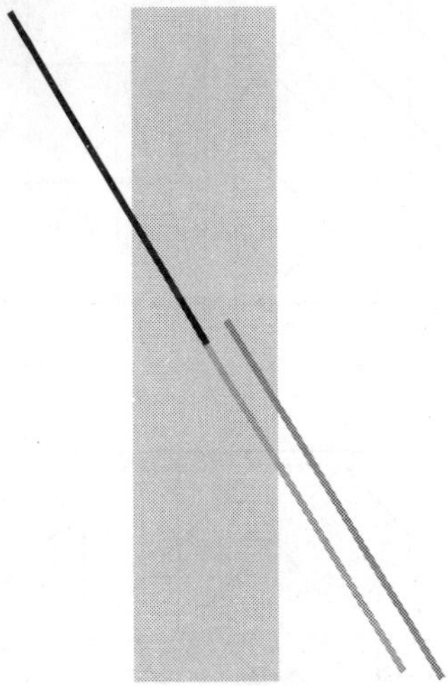

Fig. 12.12

e. Penrose stairs

It is a two-dimensional depiction of a staircase in which the stairs make four 90-degree turns as they ascend or descend yet form a continuous loop, so that a person could climb them forever and never get any higher. This is clearly impossible in three dimensions; the two-dimensional figure achieves this paradox by distorting perspective.

f. Ehrenstein illusion

The shape inside the concentric circle is a square with all sides straight.

i)

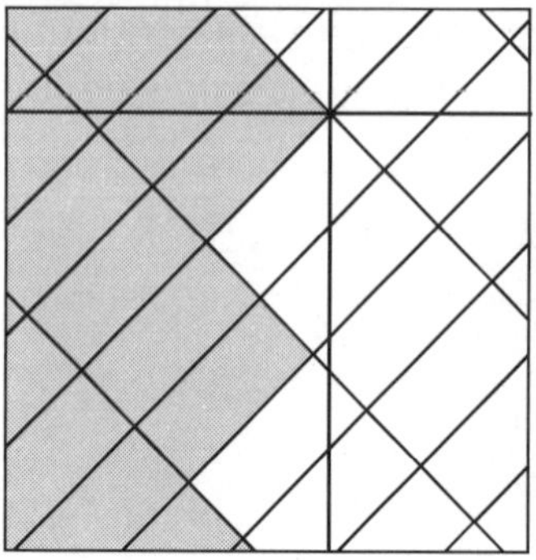

Fig. 12.13 (a)

ii)

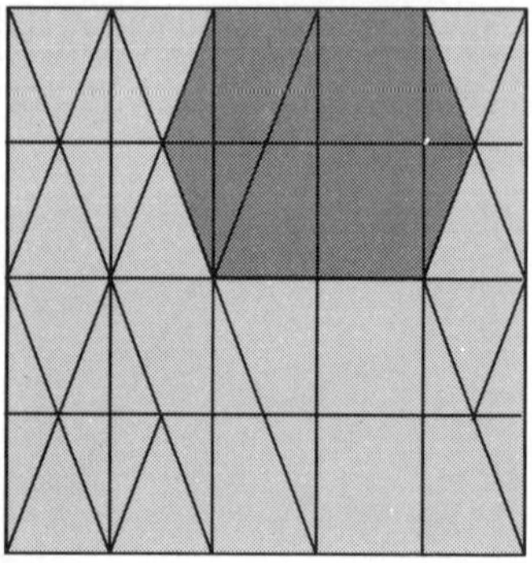

Fig. 12.13 (b)

iii)

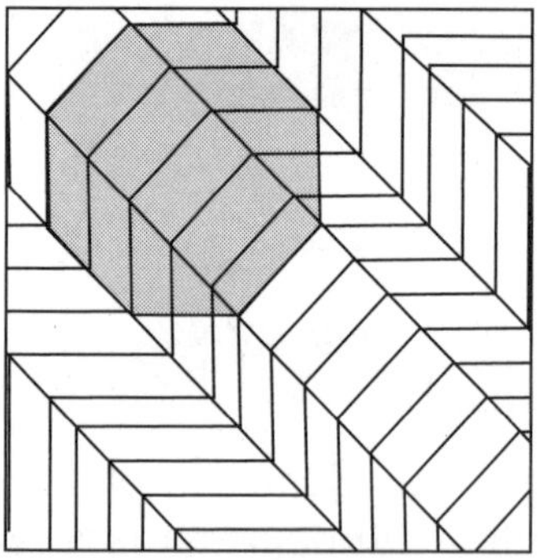

Fig. 12.13 (c)

Mathematics in everyday life (Chapter 9)

The human body is based on the Golden ratio and the number five.

1. The ratio of length of elbow to wrist to the length of wrist to fingertips is very close to phi. Ratio of the distance between the navel and knee to the distance between the knee and the end of the foot is phi.
2. The human head forms a golden rectangle with the eyes at its midpoint.

3. One full turn in a DNA molecule spiral is 34 angstroms and its width 21 angstroms; the ratio is phi.
4. We have five sense organs, five appendages in the head, arm and leg.
5. Human body is full of measurements – height, weight, quantity of water and blood, number of bones and muscles etc.
6. The concept of ratio is also very important in human body. For example the number of heart beat per minute, the ratio of height and weight (known as body mass index) etc.
7. The human body is bilaterally symmetrical.

The leaves and branches of the sneezewort plant at any stage follows the Fibonacci number pattern.

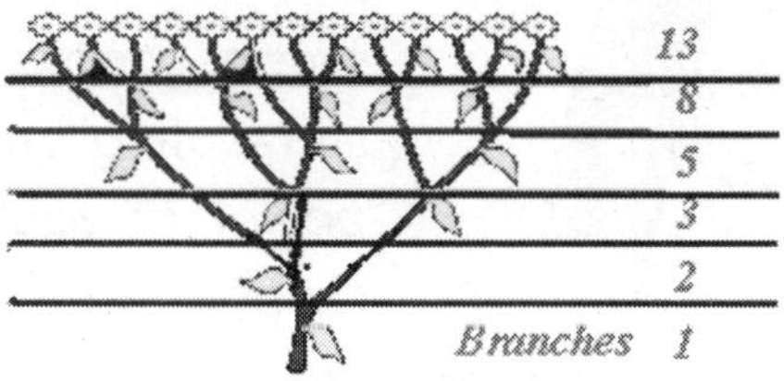

Fig. 12.14(a)

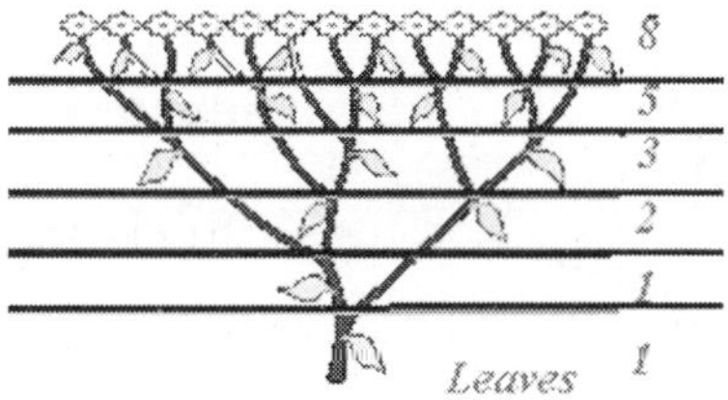

Fig. 12.14(b)

+++

New Arrivals

₹ 96/-

₹ 72/-

₹ 72/-

₹ 110/-

Visiting gymnasiums and heath clubs for maintaining physical fitness has become very much of part of our modern life. But there is little public awareness about maintaining mental health despite the fact that mental fitness is as important as physical fitness in our day–to-day life. Thanks to the lack of awareness and mental exercises, many people cut a sorry figure when it comes to problem-solving.

These books are aimed at stimulating and rejuvenating the power of your brain through a number of puzzles and brain-teasers.

How will you be benefited?

- Your brain muscles will be flexed
- Quick thinking and analytical abilities will be enhanced
- Your lateral thinking abilities will be boosted up

Now, read and restore your mental health...